ISAAC NEWTON

His Own Words

HOW to See Further

edited by Shumba

Isaac Newton - His Own Words: HOW to See Further

Published under the Shumba imprint. Part of the profit from this series of HOW books will support clean water, sanitation, and economic development work in Africa.

First Edition

Shumba is the editor and compiler of the HOW (His/Her Own Words) series of books that bring historical figures to life in their own words. Each volume draws exclusively from primary-source material, cross-checked against public-domain editions, so that you can read the actual words spoken or written rather than the paraphrase of others.

Other titles in the HOW series include: Marcus Aurelius — HOW to Master Yourself, Jesus — HOW to Live Forever, Aristotle — HOW to Find Your Purpose, Queen Elizabeth I — HOW to Reign Without a King, Queen Victoria — HOW to Reign in Grief, Nikola Tesla — HOW to Invent the Future, William Shakespeare — HOW to Tell a Story, Oscar Wilde — HOW to Master Conversation, Leonardo da Vinci — HOW to Unlock Your Genius, Florence Nightingale — HOW to Care for Others, and Mahatma Gandhi — HOW to Spin Your Own Thread.

Isaac Newton

1642 - 1727

He was born on Christmas Day 1642, in a modest Lincolnshire farmhouse, three months after his father's death — so small, his mother said, that he could have fit inside a quart pot. By three he had been left at Woolsthorpe Manor in his grandmother's care while his newly remarried mother moved a mile away; he would not fully rejoin her household for nearly a decade. Out of that fractured childhood came a mind that would rewrite the mathematics of change itself, redefine light and color, and describe, in a single law, why an apple falls and why the moon does not.

When plague closed Cambridge in 1665, sending the young Trinity College scholar home for two years, he did not idle. In what he later called his annus mirabilis, he laid the foundations of the calculus, began dismantling two-thousand-year-old assumptions about light, and worked out the mathematics that would become universal gravitation. He returned to Cambridge and built a working reflecting telescope with his own hands — so far ahead of its time it won him a Fellowship in the Royal Society before he had published a word of the theory that made him famous, and which provoked, on arrival, the first of several bitter rivalries — with Robert Hooke, later Gottfried Leibniz — that shadowed him for life. He answered each the same way: with more evidence, and more patience, than his critics could sustain.

The Principia, published in 1687 after eighteen months of near-total seclusion, unified the heavens and the earth under a single set of laws — arguably the most consequential scientific book ever written. He was, in the same lifetime, Warden and then Master of the Royal Mint, hunting counterfeiters through London's taverns and prisons; President of the Royal Society for the last quarter-century of his life; and a secret scholar of biblical prophecy. He died in London on 20 March 1727, given a state funeral in Westminster Abbey — the first scientist in England so honored. This book offers no invented deathbed wisdom, only what the record shows: a mind that refused, for sixty years, to accept an answer it had not tested.

Contents

CHAPTER 1

Method

"My Design in this Book is not to explain the Properties of Light by Hypotheses, but to propose and prove them by Reason and Experiments."
— Opticks, Book I, Part I (1704)

Isaac Newton was twenty-three when the plague closed Cambridge and sent him home to Woolsthorpe Manor for the better part of two years. He would later call this period, half in earnest and half in the flattering hindsight of an old man, his annus mirabilis — the year of wonders, when the outlines of calculus, optics, and universal gravitation all first came into view. But the habit of mind that made those years so productive was not a flash of genius so much as a discipline: a refusal to accept an explanation until it had been tested, measured, and tested again.

This is the thread that runs through everything Newton wrote, from his earliest optical papers to the last editions of the Principia he corrected by hand in his eighties. He did not trust elegant theories for their own sake. He trusted what could be shown. When his contemporaries reached for grand mechanical pictures of how the universe must work — vortices of invisible fluid sweeping the planets along, unseen springs and hooks holding matter together — Newton, more often than not, declined to guess. And he said so, in print, repeatedly, sometimes to the visible irritation of readers who wanted him to finish the story he had only half told.

It is worth pausing on how unusual this was for his time. The seventeenth century's new philosophers had already broken with the ancients; they no longer accepted that heavy things fall because it is their nature to seek the center of the earth, or that the heavens turn because they are made of a different, perfect substance than the earth below. But many of them replaced one kind of storytelling with another — mechanical stories, full of gears and fluids and collisions, that sounded scientific without being any more provable than what they replaced. Newton's great contribution to method was not that he had a better story. It was that he mostly refused to tell one at all, and insisted that a description of how something behaves is worth more than a guess about why.

That is not to say Newton had no opinions about causes. He had many, and some of them — particularly about God's role in the design of the universe — he held with total conviction. But he was unusually careful about which kind of statement he was making at any given moment: a measured fact, a mathematical deduction from that fact, or a private suspicion he could not yet prove. The chapter that follows collects Newton's own statements of method, in roughly the order he made them across his life — from the young lecturer defending his first theory of light before a skeptical Royal Society, to the old president revising, one final time, the book that had already remade the science of his age.

> *"My Design in this Book is not to explain the Properties of Light by Hypotheses, but to propose and prove them by Reason and Experiments."*
>
> — Opticks, Book I, Part I (1704)

The opening line of Newton's great treatise on light states his method before he states a single fact about light itself.

> *"Whereas the main Business of natural Philosophy is to argue from Phaenomena*

without feigning Hypotheses, and to deduce Causes from Effects, till we come to the very first Cause, which certainly is not mechanical."

— Opticks, Book III, Query 28 (4th ed., 1730)

Late in his life, revising the Opticks for what would be its final edition, Newton returned to this same principle and pushed it further — all the way, he insists, to a first cause that is not mechanical at all.

"I offer this work as the mathematical principles of natural philosophy, for the whole burden of philosophy seems to consist in this — from the phenomena of motions to investigate the forces of nature, and then from these forces to demonstrate the other phenomena."

— Preface to the Principia (1687)

In the preface to his single greatest work, Newton describes the entire enterprise of natural philosophy in two directions at once: work backward from what is observed to the forces that must be causing it, then work forward again to see if those forces predict everything else.

"I wish we could derive the rest of the phenomena of Nature by the same kind of reasoning from mechanical principles, for I am induced by many reasons to suspect that they may all depend upon certain forces by which the particles of bodies... are either mutually impelled towards one another, and cohere in regular figures, or are repelled and recede from one another. These forces being unknown, philosophers have hitherto attempted the search of Nature in vain; but I hope the principles here laid down will afford some light either to this or some truer method of philosophy."

— Preface to the Principia (1687)

Newton closes his preface with something rarer than confidence: an open admission of how much remains unknown, and a hope — not a claim — that his method might eventually reach it.

> *"To explain all nature is too difficult a task for any one man or even for any one age. 'Tis much better to do a little with certainty, & leave the rest for others that come after, than to explain all things by conjecture without making sure of any thing."*
>
> — Draft manuscript note related to the Opticks Queries

In a private working note, Newton set down what may be the plainest statement of his temperament: better a small, certain thing than a large, guessed one.

> *"[Mathematical] quantities I here consider not as consisting of indivisibles, either parts least possible or infinitely small... each time it can conveniently so be done, it is preferable to express them by finite lines visible to the eye rather than by infinitely small ones."*
>
> — Method of Fluxions (written 1671, published 1736)

Even in pure mathematics — the one field where Newton might have permitted himself abstractions no one could ever check by experiment — his instinct was toward what could be seen and measured, and away from constructs that could not.

> *"Hypotheses non fingo. I feign no hypotheses."*
>
> — General Scholium, appended to the second edition of the Principia (1713)

Four words, in the original Latin, became the most famous methodological statement in the history of science. Newton had just finished explaining that he could describe gravity mathematically — how strongly it pulls, and how that pull weakens with distance — without knowing, or claiming to know, what gravity actually is. Rather than paper over that gap with an invented mechanism, he simply declined to invent one.

> *"For whatever is not deduced from the phenomena is to be called an hypothesis; and hypotheses, whether metaphysical or physical, whether of occult qualities or mechanical, have no place in experimental philosophy."*
>
> — General Scholium, Principia (1713)

Newton goes on to explain exactly what he means by the phrase — and rules out, in one sentence, both the mystical explanations of the older philosophy and the fashionable mechanical ones that had replaced them. Experimental philosophy, in his view, answered to evidence alone.

> *"In this Philosophy Particular Propositions are inferr'd from the Phaenomena, and afterwards rendered general by Induction."*
>
> — Opticks, Book III, Query 31 (4th ed., 1730)

Late in the Opticks, Newton lays out the logical shape of his method in the plainest terms he ever used: start with what is actually observed, in particular cases, and only then — carefully, and always subject to correction — allow the conclusion to widen.

> *"And the Argument from the Phaenomena continues good till other Phaenomena occur, by which it may either be made more accurate, or liable to Exceptions."*
>
> — Opticks, Book III, Query 31 (4th ed., 1730)

Newton adds, immediately after, the clause that makes his method genuinely modern rather than merely rigorous: a conclusion drawn from the evidence remains true only until new evidence complicates it. Certainty, in his system, is always provisional, held until something better comes along to refine or overturn it.

It would be easy to mistake this caution for timidity. It was not. Newton advanced claims — about light, about gravity, about the shape of the earth — that upended a century of settled opinion, and he defended them with force when he was certain of his ground. What his method actually protected him from was not boldness but bluff: the temptation, so easy for a mind as quick as his, to fill a gap in knowledge with a plausible story and call the story an answer.

It is also worth saying plainly that Newton did not always live up to his own standard. He held private convictions — about the nature of God, about the deep history of ancient kingdoms, about the hidden workings of matter — that went well beyond what his own method could support, and he pursued some of them for decades without ever publishing a word. The Newton who insisted, in print, that he feigned no hypotheses was, in his own study, entertaining a great many. The distinction he was careful to draw was not between a mind that speculates and one that does not — no working mind manages that — but between a private hypothesis, held provisionally, and a public claim, offered as if it were proven. He kept those two boxes separate with unusual discipline, even if he did not always keep the second one empty.

This double life — rigorous in public, speculative in private — is not, on reflection, a contradiction so much as a division of labor. Newton seems to have understood that the standard he held his published work to was demanding enough that almost nothing could meet it on the first attempt. Ideas needed somewhere to live while they were still half-formed: a private notebook, a manuscript shown to no one, a letter marked confidential. What went into print, by contrast, had to have already survived years of exactly the testing described in this chapter. The chapters that follow will move back and forth between these two registers more than once — public certainty and private wondering — and it is worth remembering, each time, which box a given passage belongs in.

The chapters that follow are, in a sense, all applications of this one instinct — turned toward light, toward the heavens, toward the counterfeiters at the Mint, and eventually toward the question of God. Newton rarely changed subjects. He only ever asked, of whatever was in front of him, the same patient question: what can actually be shown?

It is worth adding a final word on why this particular figure was chosen to open a book built entirely around this one theme. Other historical subjects offer more dramatic lives, more quotable one-liners, more obviously inspirational arcs. Newton offers something rarer: a complete, documented record of a mind that built an entire philosophy around the discipline of not getting ahead of its own evidence, and then spent sixty years actually living inside that discipline, across physics, mathematics, theology, ancient history, and criminal investigation alike. Few historical figures left behind so consistent a record of the same method applied to so many different kinds of problem.

The influence of this method on everything that followed in the sciences is difficult to overstate, and worth stating plainly even so. The basic shape of what is still taught today as 'the scientific method' — observe carefully, form a hypothesis cautiously, test it against further observation, and hold the conclusion only as firmly as the evidence allows — owes an enormous debt to exactly the habits of mind collected in this chapter. Newton did not invent experimental philosophy from nothing; Galileo, Bacon, and others had cleared much of the ground before him. But it was Newton's own combination of mathematical rigor and experimental caution, demonstrated across the Opticks and the Principia alike, that became the model later scientists actually tried to imitate.

CHAPTER 2
Discovery

"In the beginning of the Year 1666 (at which time I applyed my self to the grinding of Optick glasses of other figures than Spherical,) I procured me a Triangular glass-Prisme, to try therewith the celebrated Phaenomena of Colours."

— Letter to Henry Oldenburg, 6 February 1671/72

Most of what survives in Newton's hand reads like a treatise: formal, impersonal, addressed to no one in particular. But there is one document — his first great public paper, sent to the Royal Society in February 1672 — where the voice changes. It is written in the first person, in plain narrative English, and it tells a story: a young man, a toy from a country fair, and an afternoon spent in a darkened room.

Newton was twenty-nine, newly elected a Fellow of the Royal Society on the strength of a small reflecting telescope he had built and ground himself. The Society's secretary, Henry Oldenburg, asked him to write up whatever else he had been working on. What came back overturned two thousand years of assumptions about the nature of light and color — and it did so not through abstract argument, but through the kind of experiment anyone could have tried, if anyone else had thought to try it that carefully.

The paper Newton sent to London in February 1672 was not a polished treatise prepared over years, in the manner of the Opticks that would eventually follow it by more than three decades. It was closer to a field report — an account, written soon after the fact, of what he had noticed and how he had gone about checking it. That immediacy is exactly what makes it valuable here. The formal propositions of the Opticks show Newton's conclusions, arranged and defended after the fact. This letter shows the thinking as it happened, complete with the false starts and second-guessing any real discovery involves.

It helps to picture the room itself. Newton was living in modest college rooms at Trinity, and the experiment required nothing more exotic than a shuttered window, a triangular prism bought as a curiosity rather than a scientific instrument, and a plain wall to catch the resulting light. There was no laboratory in the modern sense, no specialized apparatus beyond what a moderately curious gentleman of the period might already own. Part of what makes this letter such a good place to begin a book like this is exactly that modesty of means: the theory that eventually reorganized the science of optics began with equipment a schoolboy could have assembled, in the hands of someone willing to look more carefully than anyone had looked before.

This chapter is built largely around that one letter, published in the Philosophical Transactions as 'A Letter of Mr. Isaac Newton... containing his New Theory about Light and Colors.' It is Newton at his most readable, and it shows something the formal treatises tend to hide: that his discoveries began not with mathematics, but with curiosity about a piece of glass.

> *"Sir, To perform my late promise to you, I shall without further ceremony acquaint you, that in the beginning of the Year 1666 (at which time I applyed my self to the grinding of Optick glasses of other figures than Spherical,) I procured me a Triangular glass-Prisme, to try*

> *therewith the celebrated Phaenomena of Colours."*
>
> — Letter to Henry Oldenburg, 6 February 1671/72

The letter opens as a promise kept between two men who already knew each other — Newton had evidently told Oldenburg he had more to share, and here delivers it without preamble. The prism itself, he says plainly, was bought to try out a well-known trick, not to overturn a theory of light.

> *"And for that purpose having darkened my chamber, and made a small hole in my window-shuts, to let in a convenient quantity of the Suns light, I placed my Prisme at his entrance, that it might be thereby refracted to the opposite wall."*
>
> — Letter to Henry Oldenburg, 6 February 1671/72

This is the founding experiment of modern optics, described with the same plainness a person might use to explain how they set up a camera. A hole in a shutter. A prism. A wall.

> *"It was at first a very pleasing divertisement, to view the vivid and intense colours produced thereby; but after a while applying my self to consider them more circumspectly, I became surprised to see them in an oblong form; which, according to the received laws of Refraction, I expected should have been circular."*
>
> — Letter to Henry Oldenburg, 6 February 1671/72

The turn from pleasure to puzzlement, in a single sentence, is the moment the discovery actually begins. Everyone before Newton had watched a prism throw colors on a wall and admired the sight. Newton noticed that the shape of the patch of light did not match what the accepted theory predicted, and did not let the discrepancy go.

> *"I could scarce think, that the various Thickness of the Glass, or the termination with shadow or darkness, could have any influence on light to produce such an effect... and I doubted not, but that the Rays, after their trajection through the Prisme, moved in curve lines."*
>
> — Letter to Henry Oldenburg, 6 February 1671/72

Newton walks the reader through his own false starts before revealing the true cause — a candor rare in scientific writing of any century, where results are usually presented as though the correct answer had been obvious from the outset.

> *"I shall endeavour to testify my gratitude by communicating what my poore & solitary endeavours can effect towards ye promoting your Philosophicall designes."*
>
> — Covering note to Henry Oldenburg, 1672

Before the science comes the courtesy. Newton, still relatively unknown, thanks the Society's secretary in the modest, formal register expected of a young man addressing an established institution — a register he would abandon soon enough, once his findings began to be challenged.

What followed in the rest of the letter is a careful sequence of further experiments — passing the refracted light through a second prism, isolating single colors and refracting them again — each one designed to rule out an alternative explanation before Newton would commit to his own. Newton himself called the decisive step his experimentum crucis, a crucial experiment: rather than simply describing the colored patch on the wall, he isolated a single band of color from it, passed that single color alone through a second prism, and showed that it refracted again without splitting into further colors or changing hue. If ordinary white light were simply being modified or stained by the glass, as the received theory held, there was no reason a single color should behave any differently from the rest. That it did not — that each color, once separated, stayed exactly as it was — became the linchpin of everything that followed. It is the method described in the previous chapter, caught here in its earliest and most vivid form: not yet a set of general rules, but a young man in a dark room, unwilling to look away from a shape on a wall that did not match what he had been told to expect.

The paper made Newton famous and made him enemies in the same week. Robert Hooke, the Society's most senior experimentalist and the author of his own well-regarded book on optics, objected within days of the paper being read aloud at a Society meeting, and the argument that followed would occupy Newton, on and off, for years — resurfacing, sharpened by old resentments, in the letter examined in a later chapter of this book. Newton, stung by the reception, would later write to a friend that he regretted ever publishing at all, and for a time he genuinely withdrew from public scientific debate rather than continue defending a result he considered already proven.

It is worth noticing, too, how young the finished theory actually was when Newton sent it to London. He had performed the core experiments roughly six years earlier, during the same plague-year seclusion at Woolsthorpe that produced his first work on calculus and gravity. That he sat on the color theory for so long before publishing — turning it over privately, testing it against objections he had not yet heard — previews a pattern this book will return to more than once: Newton rarely rushed a result into print, however striking, until he had spent years quietly making sure it could survive an argument.

It is worth adding a note on Newton's actual age and standing at the moment of writing. He was not yet a famous man in February 1672 — a promising Cambridge mathematician, recently elected to the Royal Society on the strength of his telescope, but with no publication record to speak of. The confidence of the letter's opening lines, walking Oldenburg through a discovery that would eventually unseat two thousand years of received opinion about the nature of light, is striking given how little standing Newton actually had to fall back on if the argument failed to persuade. He wrote, in other words, as though the evidence alone would be enough to carry the point — which, eventually, if painfully, it was.

What belongs here, before that dispute takes over the story, is simply the discovery itself — small, domestic, and entirely his own: a hole in a shutter, a beam of sunlight, and a young man who noticed that the shape on the wall was not the shape he had been told to expect, and would not let the discrepancy go until he understood why.

CHAPTER 3
Wonder

"Do not Bodies act upon Light at a distance, and by their action bend its Rays?"
— Opticks, Book III, Query 1

When Newton finished the third book of the Opticks, he did not conclude with a summary of his results. He concluded with questions — thirty-one of them in the final edition, growing over successive printings from a modest sixteen, some of them a single sentence, others running to many pages. He called them Queries, and he wrote nearly all of them in a strange grammatical form: not 'I believe X' but 'Is it not the case that X?' — a rhetorical question, technically open, but posed by a man who plainly suspected he already knew the answer.

The Queries range across everything Newton had not yet been able to prove: the nature of heat, the cause of gravity, the structure of matter, the behavior of light at the smallest scales, and eventually — in the longest and strangest of them — the relationship between the created world and its Creator. Read together, they are less a list of hypotheses than a portrait of a mind that never stopped being curious even after it had settled the questions it could actually answer.

It is worth noticing what the Queries are not. They are not a private notebook, scribbled for Newton's own eyes and never meant to see print. He published them, revised them across four editions over more than a quarter century, and let his name stand behind every one — including the ones later science would show to be wrong. Few scientists in any era have been willing to attach their name so publicly, and for so long, to a list of things they suspected but could not yet prove.

This chapter collects a sample of the shorter Queries: Newton at his most exploratory, asking about fire, heat, and the workings of the eye with the same plain wonder he must have felt as a boy watching a prism throw colors on a wall.

> *"Do not Bodies act upon Light at a distance, and by their action bend its Rays; and is not this action (caeteris paribus) strongest at the least distance?"*
>
> — Opticks, Book III, Query 1

The very first Query, and the shortest kind: a single, precise question about whether matter can influence light without touching it.

> *"Is not Fire a Body heated so hot as to emit Light copiously? For what else is a red hot Iron than Fire? And what else is a burning Coal than red hot Wood?"*
>
> — Opticks, Book III, Query 9

Newton often reasoned by pointing at something ordinary and asking his reader to notice what was strange about it. A coal fire and a glowing iron bar, looked at closely, are the same phenomenon.

> *"Do not all fix'd Bodies, when heated beyond a certain degree, emit Light and shine; and is not this Emission perform'd by the vibrating motions of their parts? And do not all Bodies which abound with terrestrial parts... emit Light as often as those parts are sufficiently agitated?"*
>
> — Opticks, Book III, Query 8

Newton pushes the same everyday observation — that heated things glow — toward a general rule, wondering aloud whether every kind of matter might be made to shine, given enough agitation, whatever the cause of that agitation happens to be.

> *"Are not the Sun and fix'd Stars great Earths vehemently hot, whose heat is conserved by the greatness of the Bodies, and the mutual Action and Reaction between them, and the Light which they emit?"*
>
> — Opticks, Book III, Query 11

Newton wonders whether the sun and stars might simply be enormous, glowing versions of the same hot iron and burning coal discussed a few Queries earlier — vast bodies that stay hot largely because of their own size, rather than because they are made of some wholly different, celestial kind of matter.

> *"Is not the strength and vigor of the action between light and sulphureous bodies one reason why sulphureous bodies take fire more readily, and burn more vehemently, than other bodies do?"*
>
> — Opticks, Book III, Query 10

Newton reaches for a common, flammable substance — sulphur — and wonders whether its particular readiness to catch fire might be explained by some special affinity between it and light itself, an early guess at the idea that different materials interact with light in chemically distinct ways.

> *"Do not the Rays of Light in falling upon the bottom of the Eye excite Vibrations in the Tunica Retina? Which Vibrations, being propagated along the solid Fibres of the optick*

Nerves into the Brain, cause the Sense of seeing."

— Opticks, Book III, Query 12

Newton had no way to observe a nerve impulse. He guessed, correctly in its broad outline, at a mechanism for vision more than a century before anyone could test it directly.

"When a Man in the dark presses either corner of his Eye with his Finger, and turns his Eye away from his Finger, he will see a Circle of Colours like those in the Feather of a Peacock's Tail... Do not these Colours arise from such Motions excited in the bottom of the Eye by the Pressure and Motion of the Finger, as, at other times are excited there by Light for causing Vision?"

— Opticks, Book III, Query 16

A wonderfully domestic experiment — anyone can try it — used to probe a question about the physical basis of sight. Newton is, in effect, asking his reader to press on their own eye in the dark and think about what that proves.

"Are not gross Bodies and Light convertible into one another, and may not Bodies receive much of their Activity from the Particles of Light which enter their Composition?"

— Opticks, Book III, Query 30

Two and a half centuries before Einstein's $E=mc^2$ gave the idea a precise mathematical footing, Newton was already wondering, in his tentative rhetorical form, whether matter and light might be two expressions of the same underlying thing.

"Is not Vision perform'd chiefly by the Vibrations of this Medium, excited in the bottom of the Eye by the Rays of Light, and propagated

through the solid, pellucid and uniform Capillamenta of the optick Nerves into the place of Sensation?"

— Opticks, Book III, Query 21

Newton returns to the mechanism of vision a second time, from a slightly different angle, wondering whether a subtle vibrating medium — rather than the nerve fibers themselves — might be doing the actual work of carrying the sensation of light from the eye to wherever it is finally perceived.

It is easy, reading these Queries now, to score them for accuracy — some hold up remarkably well, others do not hold up at all. But that misses what makes them worth reading. Newton published his certainties as propositions, flatly stated and mathematically proved. He published his uncertainties as questions, and he published them anyway, in the same book, under his own name, for anyone to read and challenge. Very few scientists of his stature, in any era, have been willing to put their open questions on permanent public record beside their closed ones.

Notice, too, the direction his wonder tends to run. Newton is rarely content to observe a single striking fact and leave it there; a burning coal, a red-hot iron bar, and the light of the sun itself are, by the end of a few short Queries, drawn together into a single suspected pattern. That reflex — noticing that two unrelated-looking things might in fact be the same thing, seen at different scales — is close to the center of everything he accomplished, and it shows up here in its rawest, least-proven form, well before it hardens into mathematics.

There is also something worth noticing in the sheer range of subjects the short Queries touch in quick succession: fire, sulphur, the eye, the ear, the retina, the structure of a peacock feather. Newton did not confine his curiosity to the grand, cosmic questions that dominate the later, longer Queries. He was equally willing to wonder about a burning match or the color in a bird's tail, and he treated both kinds of question — the vast and the small — with exactly the same seriousness, never signaling that one was more worthy of his attention than the other.

It is worth pausing, too, on how often Newton's wonder returns specifically to the eye. Across this short sample alone, he asks about the retina, the pressure-induced colors seen with a closed eyelid, and now the vibrating medium of vision itself — three separate Queries, spread across different editions of the book, all circling the same unresolved question of how a beam of light becomes a seen thing inside a living creature. It is tempting to read something personal into this persistence: a man who spent his working life training his own eyes on prisms, telescopes, and manuscripts, still wondering, decades on, exactly how the instrument he relied on most actually did its work.

The habit reveals something about how he actually worked. Wonder, for Newton, was not a stage that ended once the method described in the first chapter took over. It ran alongside it, for his entire life, in a parallel column of things he had noticed but not yet proved — and he never stopped adding to the list.

CHAPTER 4

Standing on Others

"If I have seen further it is by standing on ye shoulders of Giants."

— Letter to Robert Hooke, 5 February 1675/76

No sentence Newton ever wrote is more famous, or more consistently misread. Four years after his prism letter made him a name at the Royal Society, Newton found himself in a bitter dispute with Robert Hooke — the Society's most senior experimentalist, a man who had built much of his own reputation on optics and who resented, sharply, a younger rival's rapid rise. The two had already clashed once, publicly, over Newton's theory of color. In early 1676 Newton wrote to Hooke privately, proposing that whatever disagreements remained between them be settled as 'a Consultation' between friends rather than fought out again in front of the whole Society.

It is in that conciliatory letter that the famous line appears. Read in isolation, on a coffee mug or a graduation card, it sounds like pure humility: a great man crediting his forebears. Read in the letter it actually came from, it is more complicated than that — and the complications are worth knowing, because they say something true about how credit and rivalry actually worked between two brilliant, prickly men who never much liked each other.

Robert Hooke was noted, unkindly, by contemporaries for his stooped and slight physical stature. Historians have long debated whether Newton's choice of image — dwarfs seeing further by standing on the shoulders of giants, an idea itself borrowed from a twelfth-century philosopher named Bernard of Chartres — was entirely innocent. What is certain is that the peace the letter proposed did not last. Hooke never acknowledged receiving it, so far as any surviving record shows, and the two men remained rivals for the rest of Hooke's life.

The dispute had a history by the time this letter was written. Hooke had published his own influential book on optics, the Micrographia, in 1665, and when Newton's 1672 paper on the theory of light and color was read before the Royal Society, Hooke stood up within days to object that Newton had contributed nothing Hooke's own book had not already covered. Newton, by every account touchy about his priority in a way that never really softened with age, took the challenge badly. The 1676 letter, with its famous line, was written in the middle of this longer, unresolved argument — not as an isolated gesture of goodwill, but as one move in a running match between two men who had already been sparring in print for years.

> *"What's done before many witnesses is seldome without some further concern than that for truth: but what passes between friends in private usually deserves the name of consultation rather than contest, & so I hope it will prove between you & me."*
>
> — Letter to Robert Hooke, 5 February 1675/76

Newton opens by proposing a private peace — an argument fought only for truth's sake, he suggests, belongs behind closed doors, not before an audience.

> *"What Des-Cartes did was a good step. You have added much several ways, & especially in taking ye colours of thin plates into philosophical consideration."*
>
> — Letter to Robert Hooke, 5 February 1675/76

Before the famous line, Newton credits two specific debts: to Descartes, and to Hooke himself, for work on the colors seen in thin films of oil or soap.

> *"I am extremely glad therefore that in my late papers I stated my new Theory of light and colours after such a manner, as might occasion those things to be examined, which might have been passed by without regard, had they been delivered only by hints."*
>
> — Letter to Robert Hooke, 5 February 1675/76

In an unusually generous aside, Newton suggests that publishing his theory in full, exposed detail — rather than the more guarded, hinting style some of his contemporaries preferred — was itself a service to Hooke and the others, since it gave them something concrete enough to actually test and improve on.

> *"If I have seen further it is by standing on ye shoulders of Giants."*
>
> — Letter to Robert Hooke, 5 February 1675/76

The sentence that followed, offered as the conclusion to that short list of debts — and which has outlived every other line either man ever wrote.

> *"But I make no question but you have divers very considerable experiments besides those you have published, & some it's very probable the same wth some of those in my late papers. Two at least there are wch I know you have observed."*
>
> — Letter to Robert Hooke, 5 February 1675/76

The letter continues past the famous line into something closer to its real business — a careful, almost wary accounting of who had observed what first.

It would be a disservice to reduce this to a simple story of insult dressed as flattery; the truth sits somewhere less tidy. Newton's debts to Descartes and to Hooke, named in the same paragraph, were real ones, and the desire to keep a dispute out of public view was, on its own terms, a reasonable thing to want. But it would be an equal disservice to sand the episode down into a greeting-card sentiment about humility, when the men involved were locked in a rivalry that neither ever really let go of. Newton, decades later, as President of the Royal Society, would be accused by Hooke's allies of erasing the older man's portrait and instruments from the Society's rooms after Hooke's death — an accusation impossible to verify fully, but consistent with everything else known about how long Newton could hold a grudge.

The two men never reconciled in any way the record shows. Hooke continued to needle Newton over the theory of light for years, and Newton, for his part, waited — with what looks in hindsight like real patience — until Hooke's death in 1703 before publishing the Opticks at all, sparing himself, deliberately or not, one more round of the argument with the one rival most likely to relish it. Whatever else the 'shoulders of giants' letter was, it was not the end of the story between them; it was, at best, a truce that held for exactly as long as it took the next dispute to arrive.

There is a further wrinkle worth knowing. No authentic portrait of Robert Hooke survives today, and the Royal Society's own collection lost whatever likeness it once held sometime after Hooke's death — a gap historians have long found suspicious given Newton's documented dislike of the man and his long tenure as the Society's president in the years the portrait went missing. No direct evidence proves Newton was responsible, and this book will not claim more certainty than the record supports. But the absence itself has become, fairly or not, part of how the rivalry is remembered: even the physical trace of Hooke's face did not survive contact with Newton's long institutional shadow.

It is worth remembering, too, that Hooke was not simply an obstacle in Newton's path. He was, by most accounts, one of the most versatile experimental minds of the entire century — credited with foundational work on elasticity, microscopy, and the mechanics of springs, and the first to publish detailed, beautifully illustrated observations of the microscopic world in his own Micrographia. That a man of this stature felt so consistently overshadowed and provoked by Newton says something about the sheer scale of what Newton was producing during these same years, even as it says something less flattering about how Newton chose to treat a genuine rival.

What survives, in the end, is the sentence itself, and it is worth taking seriously on its own account, independent of the tangled motives behind it. Newton did understand, and say plainly, that no one sees anything in science alone. He named the men he had climbed on. That the climbing did not stop him from also elbowing his fellow climbers is simply the rest of the truth — and a truer picture of how discovery actually happens than the sentence alone would suggest.

CHAPTER 5

Faith and First Causes

"When I wrote my treatise about our System, I had an eye upon such Principles as might work with considering men for the belief of a Deity."

— First letter to Richard Bentley, 10 December 1692

In 1692, a young clergyman named Richard Bentley was preparing the inaugural set of lectures under a new endowment, established by the will of the scientist Robert Boyle, meant to use the new natural philosophy to argue against atheism. Bentley had read the Principia — or tried to; it was, and remains, a formidably difficult book — and he wrote to its author directly, asking whether Newton's own discoveries about gravity could be used to support the case for a Creator.

What followed was a short correspondence, four letters from Newton across a little more than a year, that gives the clearest picture available of how Newton himself connected his mathematics to his faith. He was careful, as always, to separate what he could prove from what he merely believed — but on the question of whether the universe pointed toward a designing intelligence, he did not hold back.

This chapter follows the first half of that exchange: Newton establishing, in his opening letter, why he had written the Principia with theology already in mind, and then, pressed by Bentley for specifics, drawing a careful line around what his own science could and could not claim to prove about God.

It is worth knowing a little about Bentley himself, since the letters were shaped by who was asking. Richard Bentley was not a scientist but a classical scholar, still in his twenties, chosen to deliver the first set of lectures under a bequest left by Robert Boyle specifically to defend Christianity against 'notorious infidels' using the tools of the new science. Bentley needed Newton's authority behind his argument, and he was not shy about pressing for it. The correspondence that survives shows a careful, sometimes cautious physicist being pulled, letter by letter, toward stating his religious convictions more plainly than he had ever done in print.

> *"When I wrote my treatise about our System, I had an eye upon such Principles as might work with considering men for the belief of a Deity, and nothing can rejoice me more than to find it useful for that purpose. But if I have done the public any service this way, it is due to nothing but industry and a patient thought."*
>
> — First letter to Richard Bentley, 10 December 1692

Newton opens the correspondence with something close to a confession of intent: the Principia was not written as neutral mathematics that happened to have religious implications. The implications were part of the design from the start.

> *"The same power whether natural or supernatural, which placed the Sun in the center of the orbs of the six primary Planets, placed Saturn in the center of the orbs of his five secondary Planets & Jupiter in the center of the orbs of his four secondary ones & the earth in the center of the Moons orb."*
>
> — First letter to Richard Bentley, 10 December 1692

Newton points to the tidy, repeated pattern of the solar system — planet after planet with its moons arranged around it in nested order — as the kind of coincidence he did not believe was accidental.

> *"So then gravity may put ye planets into motion but without ye divine power it could never put them into such a Circulating motion as they have about ye Sun, & therefore for this as well as other reasons I am compelled to ascribe ye frame of this Systeme to an intelligent Agent."*
>
> — First letter to Richard Bentley, 10 December 1692

Newton draws the distinction with unusual clarity: gravity, on its own mathematics, explains how bodies pull toward one another, but not why they should have started out moving sideways in the first place, in the neat, roughly circular orbits actually observed.

> *"But how the matter should divide it self into two sorts... I do not think explicable by mere natural causes but am forced to ascribe it to the counsel & contrivance of a voluntary Agent."*
>
> — First letter to Richard Bentley, 10 December 1692

Discussing why some matter in the universe became luminous (forming stars) while the rest remained dark (forming planets), Newton reaches the limit of what his physics can explain and names, deliberately, what he believes lies past that limit.

> *"You sometimes speak of gravity as essential & inherent to matter: pray do not ascribe that notion to me, for the cause of gravity is what I do not pretend to know, & therefore would take more time to consider of it."*
>
> — Second letter to Richard Bentley, 17 January 1692/93

When Bentley's draft lectures seemed to attribute a stronger claim to Newton than he had actually made, Newton corrected him immediately and precisely — a small but telling demonstration of the same discipline described in the opening chapter, applied here to his own reputation.

It would be easy to read this exchange as a scientist bending his results to fit a predetermined religious conclusion. Newton would have rejected that reading, and the letters themselves resist it: notice how carefully he distinguishes, in the second letter, between what gravity does — which he can describe with total mathematical precision — and what gravity is, which he refuses to claim to know even when it would have strengthened Bentley's argument to say so. The design argument he offers is built entirely from gaps he considered genuine: places where his own physics, honestly examined, ran out of explanatory power.

It is also worth noticing what Newton does not do in these letters. He does not reach for miracles to explain ordinary planetary motion, which he had just spent the Principia demonstrating followed fixed, discoverable laws. His argument is narrower and, in a sense, more modest: not that God intervenes moment to moment to keep the planets moving, but that the initial arrangement — the tidy, repeated pattern of planets and their moons, all turning the same way, all roughly in the same plane — is not the kind of thing blind mechanical causes would be expected to produce on their own. It is an argument from the setup of the system, not from its ongoing operation.

It is also worth noting that Bentley's lectures, once delivered, became genuinely influential — one of the founding texts of what later scholars would call physico-theology, the attempt to argue for God's existence from the evidence of natural science rather than from Scripture or tradition alone. Newton's correspondence supplied much of the scientific backbone for an argument that would go on to shape English religious thought for the better part of a century, well beyond anything Bentley could have achieved from the pulpit alone.

It is worth remembering, too, the position Newton actually occupied when Bentley first wrote to him. The Principia had been published only five years earlier, and its reception, though eventually triumphant, had been genuinely contested — continental philosophers were still working out whether to accept a force that acted, apparently, across empty space with no visible mechanism behind it. Bentley's request arrived, in other words, at a moment when Newton's own scientific authority was still being tested in public, and his willingness to lend that still-uncertain authority to a young clergyman's theological project says something about how confident he already was, privately, that his physics and his faith belonged together.

Whether that argument still convinces a reader today is a separate question from whether Newton meant it sincerely. Every piece of surviving evidence suggests he did — and that the mathematics he is remembered for and the faith he is not were, in his own mind, never two separate projects.

CHAPTER 6
Providence and Gravity

"Gravity must be caused by an agent acting constantly according to certain laws, but whether this agent be material or immaterial is a question I have left to the consideration of my readers."

— Fourth letter to Richard Bentley, 1693

The correspondence with Richard Bentley did not end with the careful distinctions of the second letter. Over two more letters, spanning into 1693, Newton pushed further into a question his own Principia had opened but deliberately not answered: if gravity is not a property built into matter itself — and Newton was adamant that it could not be — then what, or who, keeps it acting, everywhere, all the time, according to the same unbending law?

This is Newton at his most philosophically exposed. The mathematics of the Principia describes gravity's behavior with a precision no one had ever achieved for any force in nature. But behavior is not cause, and Newton knew the difference better than anyone alive. In these final two letters to Bentley, he lets the reader see what he thought was on the other side of that gap.

It is easy, from three centuries' distance, to forget how strange Newton's own achievement looked to his contemporaries. He had shown that the same force pulling an apple to the ground also held the moon in its orbit and the planets in theirs — a single law spanning distances from a garden to the edge of the known solar system. To many of his own readers, that alone felt uncomfortably close to magic: an invisible influence, reaching across empty space with no visible mechanism, no chain of gears or fluid to carry it. Newton's own unease with that picture, expressed candidly in these letters, was not false modesty. He shared, at least in part, his critics' suspicion that something about the story was still missing.

Continental philosophers, Leibniz chief among them, pressed exactly this objection for years: that Newton's gravity, whatever its mathematical elegance, was a return to precisely the kind of unexplained 'occult quality' the new philosophy had prided itself on banishing from science. Newton's letters to Bentley are, in part, a private rehearsal of the answer he would eventually need for that public criticism — an answer that amounted, in the end, to conceding the point about mechanism while refusing to concede that the mathematics was therefore worthless.

> *"It is inconceivable that inanimate brute matter should, without the mediation of something else which is not material, operate upon and affect other matter without mutual contact."*
>
> — Third letter to Richard Bentley, 25 February 1692/93

Newton states his core objection to treating gravity as simply built into matter: two bodies with no medium between them, pulling on each other across empty space, is not something dead matter can do on its own, in his view — something else must be involved.

> *"That gravity should be innate, inherent, and essential to matter, so that one body may act upon another at a distance through a vacuum... is to me so great an absurdity, that I believe no man who has in philosophical matters a competent faculty of thinking can ever fall into it."*
>
> — Third letter to Richard Bentley, 25 February 1692/93

As strong a statement of conviction as Newton ever put in writing — precisely on the question his mathematics could describe but not settle.

> *"Gravity must be caused by an agent acting constantly according to certain laws, but whether this agent be material or immaterial is a question I have left to the consideration of my readers."*
>
> — Fourth letter to Richard Bentley, 1693

The final word of the exchange leaves the door open rather than closing it. Even here, at the edge of theology, Newton stops short of stating as fact what he cannot demonstrate — he offers his readers the question, not a forced conclusion.

> *"If all the matter of the universe were evenly scattered throughout all the heavens, and every particle had an innate gravity towards all the rest... some of it would convene into one mass and some into another, so as to make an infinite number of great masses, scattered at great distances from one to another throughout all that infinite space. And thus might the sun and fixed stars be formed."*
>
> — First letter to Richard Bentley, 10 December 1692

One of the earliest serious attempts, by anyone, to reason mathematically about how the large-scale structure of the universe — stars clumped and scattered rather than smeared evenly through space — might have come to look the way it does.

> *"But if the matter was evenly disposed throughout an infinite space, it could never convene into one mass; but some of it would convene into one mass and some into another,*

so as to make an infinite number of great masses."

— First letter to Richard Bentley, 10 December 1692

Newton returns to the same reasoning a second time in the same letter, testing it from a slightly different angle — a habit visible throughout his correspondence of circling back to check a conclusion rather than moving straight past it once stated.

Later physicists would eventually offer their own answers to the question Newton left open — general relativity describes gravity not as a force transmitted by an agent at all, but as the curvature of space and time itself, a picture Newton had no way to imagine. It would be unfair to expect him to have reached it. What is fair to notice is the shape of the question he was willing to sit with, unresolved, for the rest of his working life: he built the mathematics that predicted the tides, the orbits of comets, and the shape of the earth with total confidence, and then, at the very center of that achievement, told his readers plainly that the deepest cause of it all was a mystery he had chosen not to pretend to solve.

It is worth setting the third letter's forceful language beside the fourth letter's open question and letting the contrast stand. In the space of a single ongoing correspondence, Newton could declare one possibility 'so great an absurdity' that no competent thinker could accept it, and then, a few months later, decline to say which of the remaining possibilities was true. That combination — total confidence about what could be ruled out, paired with real humility about what remained — is, in miniature, the whole temperament this book has tried to trace across every subject he ever took up.

It is worth adding, finally, how unusual it was for a natural philosopher of Newton's stature to leave a question this central so visibly open. The Principia had made him arguably the most authoritative scientific voice in Europe; a less careful man, having achieved that kind of standing, might easily have been tempted to fill the remaining gap with a confident guess, secure in the knowledge that few readers were equipped to challenge him. Newton had exactly that opportunity, in these letters to Bentley, and did not take it. The gap between what gravity does and what gravity is remained, in his own published and private writing alike, exactly the size his evidence actually supported — no larger, and no smaller.

CHAPTER 7
Color

"The Light of the Sun consists of Rays differently Refrangible."
— Opticks, Book I, Proposition II

Before Newton, the leading theory of color held that white light was pure and simple, and that colors were produced by modifying it — darkening it, in various ways, as it passed through glass or reflected off a surface. Newton's prism experiments overturned this completely, and the case he built for the new view, laid out across the opening propositions of the Opticks, remains one of the clearest pieces of experimental reasoning in the history of science: patient, layered, and designed at every step to rule out the alternative before claiming the point.

This chapter draws from that formal argument, told in Newton's own careful, exacting prose — the same discovery narrated in his letter to Oldenburg, now built into rigorous, repeatable proof. Where the earlier letter shows the moment of noticing something strange, the Opticks shows the years of careful work that followed: variation after variation on the same basic setup, each one designed to close off a different possible objection before a skeptical reader could raise it.

The structure Newton settled on — a numbered sequence of Propositions, each following logically from the Definitions and Axioms and from the Propositions before it — was borrowed deliberately from geometry, the one field of knowledge whose conclusions no one in his century seriously disputed. By building his case for a new theory of color in the same form as a proof in Euclid, Newton was making an argument about the argument itself: that a claim about the physical world, tested carefully enough, could carry the same weight as a mathematical certainty.

It is worth noticing, too, how much of the book's persuasive power comes from redundancy rather than any single knockout demonstration. Newton does not offer one clever experiment and declare the case closed. He offers the same basic result approached from several different directions — through refraction, through reflection, through recombination — so that a skeptical reader unconvinced by one line of evidence is confronted with three or four more, each independently arriving at the same conclusion. It is persuasion by accumulation as much as by any single decisive stroke.

> *"I took a black oblong stiff Paper... One of these parts I painted with a red colour and the other with a blue. The Paper was very black, and the Colours intense and thickly laid on, that the Phaenomenon might be more conspicuous."*
>
> — Opticks, Book I, Part I, Experiment 1

Newton's experiments are, before anything else, physically vivid — described in enough concrete detail that a reader three centuries later could still assemble the same paper, prism, and window shutter and watch the same thing happen.

> *"I found that if the refracting Angle of the Prism be turned upwards... its blue half will be lifted higher by the Refraction than its red half."*
>
> — Opticks, Book I, Part I, Experiment 1

The core observation, stated plainly: blue light bends more than red light does, passing through the same glass at the same angle. Different colors, it turns out, are not modifications of the same light — they are physically distinct from one another.

> *"This Image or Spectrum was coloured, being red at its least refracted end, and violet at its most refracted end, and yellow, green and blue in the intermediate Spaces."*
>
> — Opticks, Book I, Part I, Experiment 3

The now-familiar order of the spectrum, described for what may have been the first time as an ordered physical fact rather than a decorative effect.

> *"The Light of the Sun consists of Rays differently Refrangible."*
>
> — Opticks, Book I, Proposition II

The conclusion, stated as a formal theorem after the experiments that prove it: sunlight, which looks perfectly white and simple to the naked eye, is in fact a mixture of every color, each bending by its own fixed and different amount.

> *"Homogeneal Light is refracted regularly without any Dilatation splitting or shattering of the Rays."*
>
> — Opticks, Book I, Proposition V

Having separated the colors once, Newton shows that a single isolated color, passed through a second prism, does not split further or change — proof that each color is a stable, fundamental component of white light rather than an accident of the first prism.

> *"The same Colours in Specie with these Primary ones may also be produced by Composition: For, a mixture of Yellow and Blue makes Green... and a Composition of all the Primary Colours in due Proportion will produce perfect White."*
>
> — Opticks, Book I, Part II, Proposition VI

Newton closes the loop the other direction — not just separating white light into its component colors, but reassembling those same colors back into white, showing that the process works cleanly in both directions.

> *"Colours in the Object are nothing but a Disposition to reflect this or that sort of Rays more copiously than the rest."*
>
> — Opticks, Book I, Part II, Proposition II

Newton reaches the conclusion's most counterintuitive implication directly: an apple is not itself red. It is a surface disposed, by its physical structure, to reflect red-making rays more strongly than the others already present in ordinary white light — color, in other words, belongs to the interaction, not to the object alone.

> *"The Species of Colours, and Degrees of Refrangibility proper to any particular sort of Rays, are not mutable by Refraction, nor by Reflexion from natural Bodies, nor by any other Cause, that I could yet observe."*
>
> — Opticks, Book I, Part II, Proposition II

Newton generalizes the conclusion once more, ruling out every alternative cause he could think to test — reflection, refraction, or any other process — that might have changed a ray's color, and finding none of them able to do it.

It is worth remembering how radical this was received to be. Newton's paper did not simply add a detail to the existing theory of light; it replaced the theory's central assumption — that white was the pure, simple starting point, and colors were degradations of it. Newton showed the picture was backwards: white is the compound, and the separate colors are the fundamental, unchanging things.

That the case survived every challenge — including a bruising public dispute with Robert Hooke that opens the next stage of this story — says as much about the strength of Newton's experimental method as it does about the color itself. Two hundred years later, when physicists began building instruments far more sensitive than any seventeenth-century prism, Newton's account of the spectrum held up essentially unchanged. Few first papers, in any field, have aged so well.

It is worth adding one further point of context: the Opticks did not appear until 1704, more than three decades after the experiments it describes were first performed. Newton spent those decades revisiting, refining, and re-testing the same basic result rather than rushing it into print — a pattern this book examines more directly in a later chapter, but visible here already in the sheer polish and redundancy of the finished argument.

It is worth noting, too, how thoroughly Newton's account displaced not just the specific theory it argued against, but an entire older way of talking about color. Painters, dyers, and glassmakers had for centuries treated color as a kind of substance — something added to or subtracted from an object, like a pigment mixed into paint. Newton's proposition above reframes color as a disposition, a tendency to interact with light in a particular way rather than a thing possessed outright. That shift in vocabulary, as much as any single experiment, is part of what made his account genuinely difficult for contemporaries to absorb: it asked them to stop thinking of color as a property of objects at all.

CHAPTER 8

Proof

"The Angle of Reflexion is equal to the Angle of Incidence."

— Opticks, Book I, Axiom II

Newton opened the Opticks not with an experiment but with a short, austere list: eight Definitions and eight Axioms, laid out in the manner of a geometry textbook, before a single ray of colored light appears on the page. It is easy to skip past this section, and most modern readers do. But it reveals something essential about how Newton believed knowledge should be built — from the ground up, one exactly stated term at a time, so that nothing later in the book could rest on a word whose meaning had been left to drift.

This chapter draws from that foundation, alongside the Principia's own account of how forces should be inferred from motion. Together they form Newton's closest thing to a formal statement of what counts, in his view, as proof.

Newton's model for this kind of structure was Euclid's Elements, the geometry textbook every educated reader of his century would have studied as a matter of course: a small number of definitions and self-evident axioms, followed by a long chain of propositions, each one resting only on what had already been established. By opening the Opticks the same way, Newton was quietly asking his readers to hold a book about physical light — smells, colors, glass, sunlight through a window — to the same standard of certainty they already granted to a book about abstract triangles and circles. It was, in its way, as bold a claim as anything in the book's later chapters.

"By the Rays of Light I understand its least Parts, and those as well Successive in the same Lines, as Contemporary in several Lines."

— Opticks, Book I, Definition I

Before making any claim about light, Newton first defines, with almost legalistic care, exactly what he means by the word 'ray' — a discipline that runs through everything he wrote.

"Refrangibility of the Rays of Light, is their Disposition to be refracted or turned out of their Way in passing out of one transparent Body or Medium into another."

— Opticks, Book I, Definition VII

A second key term, defined with the same exactness — the property of bending differently that would, a few Propositions later, turn out to be the whole secret of the spectrum.

"The Angle of Reflexion is equal to the Angle of Incidence."

— Opticks, Book I, Axiom II

A law known since antiquity, restated here as a foundation stone — something so well established that Newton could build every later, novel claim on top of it without needing to reprove it each time.

"Wherever the Rays which come from all the Points of any Object meet again in so many Points after they have been made to converge by Reflection or Refraction, there they will make a Picture of the Object upon any white Body on which they fall."

— Opticks, Book I, Axiom VII

Newton uses this principle to explain, among other things, how the eye forms an image on the retina — grounding an explanation of vision itself in the same small set of geometrical rules used to explain a telescope.

> *"Refraction out of the rarer Medium into the denser, is made towards the Perpendicular; that is, so that the Angle of Refraction be less than the Angle of Incidence."*
>
> — Opticks, Book I, Axiom IV

Another inherited rule, stated without ceremony — light bends toward the perpendicular when it enters a denser medium, a fact well known to lens-makers long before Newton, here simply given its exact, checkable form.

The habit of defining terms before using them, and of establishing simple, well-tested rules before building anything ambitious on top of them, is easy to underrate because it produces no dramatic discoveries by itself. But it is the scaffolding that made Newton's dramatic discoveries trustworthy rather than merely striking. A claim built on carefully defined terms and previously established axioms can be checked, argument by argument, by anyone willing to do the work — which is, in the end, the entire difference between a proof and an opinion.

This is, in a real sense, the least glamorous chapter in this book, and deliberately so. Newton's fame rests on his conclusions — the spectrum, the law of gravity, the calculus. But the conclusions were only as strong as the eight quiet Definitions and eight quiet Axioms sitting beneath them, unread by most people who have heard of the book they open. A proof, Newton understood better than almost anyone, is only as trustworthy as its weakest assumption, and he made sure his own assumptions were stated plainly enough that a reader could check every one.

It is worth noticing what Newton chose to state as an Axiom rather than prove from scratch. The law of reflection, the behavior of light entering a denser medium — these were not Newton's discoveries. They were centuries old, tested and retested by generations of lens-grinders and astronomers before him, and Newton was content to borrow them without ceremony, exactly as a modern scientist borrows the value of a physical constant measured by someone else. Original work, in his method, meant building carefully on what had already been secured, not reproving everything from nothing each time a new question arose.

It is worth adding, finally, a word about the Principia's own version of this same structure, since the Opticks was not Newton's only attempt at it. The Principia opens with its own Definitions and three Axioms — famously, the three laws of motion — before a single planetary orbit is calculated. The pattern is identical across both of Newton's major works: settle the vocabulary, settle the ground rules, and only then attempt the harder claims. That he built two entirely different branches of science, decades apart, on the same underlying architecture suggests this was not a stylistic habit borrowed from geometry for one particular book, but a genuine conviction about how any reliable knowledge, in any field, had to be assembled.

CHAPTER 9

Questions Without Answers

"Doth not the Refraction of Light proceed from the different density of this aethereal Medium in different places?"

— Opticks, Book III, Query 19

The Queries introduced in an earlier chapter grew, across the different editions of the Opticks, from a handful of short questions into a much longer set touching heat, gravity, electricity, and the fine structure of matter. This chapter turns to that second tier: questions Newton posed about the physical world that reached beyond what his own century's instruments could test, some of which anticipated ideas that would not be properly worked out for two hundred years.

Read as a group, they show a mind willing to leave a permanent, published record of its own uncertainty — not tucked away in a private notebook, but printed, under his name, in the same volume as his proven results.

It helps to remember what instruments Newton actually had available. There was no way, in his lifetime, to observe an atom, measure the speed of a nerve impulse, or detect a gravitational wave. Every question in this chapter was reasoned toward from indirect evidence — the length of a shadow, the behavior of a compass needle, the steady, undiminished motion of the planets over the whole span of recorded astronomy — pushed as far as inference could carry it, and then left standing as a question once the evidence ran out.

"Do not Bodies and Light act mutually upon one another; that is to say, Bodies upon Light in

> *emitting, reflecting, refracting and inflecting it, and Light upon Bodies for heating them, and putting their parts into a vibrating motion wherein heat consists?"*
>
> — Opticks, Book III, Query 5

Newton proposes, tentatively, that heat itself might be nothing more than matter vibrating — a guess that anticipates, in rough outline, the modern kinetic theory of heat.

> *"Doth not the Refraction of Light proceed from the different density of this aethereal Medium in different places, the Light receding always from the denser parts of the Medium?"*
>
> — Opticks, Book III, Query 19

Newton's proposed 'aethereal Medium' — an invisible substance filling all space — turned out not to exist. But the underlying instinct, that refraction has something to do with a change in the medium light travels through, was sound.

> *"May not Planets and Comets, and all gross Bodies, perform their Motions more freely, and with less resistance in this aethereal Medium than in any Fluid... And may not its resistance be so small, as to be inconsiderable?"*
>
> — Opticks, Book III, Query 22

Newton is here trying to explain why the planets, moving through whatever fills space, do not visibly slow down over recorded history — reasoning his way toward something close to the modern idea of near-empty space, even while still assuming a medium filled it.

> *"Are not gross Bodies and Light convertible into one another, and may not Bodies receive much of their Activity from the Particles of Light which enter their Composition?"*

— Opticks, Book III, Query 30

Repeated here from an earlier chapter because it belongs in both places: a wonder at the boundary between light and matter, and a genuinely open scientific question Newton had no way to close.

> *"Have not the small Particles of Bodies certain Powers, Virtues, or Forces, by which they act at a distance... For it's well known, that Bodies act one upon another by the Attractions of Gravity, Magnetism, and Electricity."*
>
> — Opticks, Book III, Query 31

The opening of the longest Query in the book, in which Newton wonders whether the same kind of invisible, action-at-a-distance force he had already proven for gravity might also govern how the smallest particles of matter cohere, dissolve, and combine with one another.

> *"How these Attractions may be perform'd, I do not here consider. What I call Attraction may be perform'd by impulse, or by some other means unknown to me. I use that Word here to signify only in general any Force by which Bodies tend towards one another, whatsoever be the Cause."*
>
> — Opticks, Book III, Query 31

Newton pauses, in the middle of his own speculation, to guard against being misread — insisting that the word 'attraction' is a placeholder for an unknown mechanism, not a claim about what that mechanism actually is. The same caution that shaped his letters to Bentley about gravity reappears here, applied to matter at the smallest scale.

> *"Now by the help of these Principles, all material Things seem to have been composed of the hard and solid Particles above-mention'd,*

variously associated in the first Creation by the Counsel of an intelligent Agent."

— Opticks, Book III, Query 31

Query 31 closes, after many pages of chemical speculation, by circling back to the same theological ground explored elsewhere in this book — the smallest particles of matter, like the largest arrangement of planets, ultimately traced back to a deliberate, ordering cause rather than blind chance.

Some of these questions were answered, eventually, by other people, using instruments and mathematics Newton never had access to. Others — the aether chief among them — turned out to be dead ends, discarded once better evidence arrived. Newton could not have known, in advance, which of his questions belonged to which category. What he did know was that a question honestly asked and left open was worth more than a guess dressed up as an answer, and he was willing to let history judge which of his guesses had been right.

It is worth adding that Newton was rarely satisfied simply to wonder in isolation. Query 31, quoted above, would go on to become one of the most influential single passages he ever wrote — not for its answer, which it does not give, but for the question itself. Chemists across the following century built entire tables of 'affinities' between substances trying to work out exactly which bodies attracted which, directly inspired by Newton's speculation that the same force ruling the heavens might rule the smallest particles of matter too. A question, left carefully enough on the record, can do as much work in the world as an answer.

Notice, too, how naturally the chapter's final quote returns to territory covered several chapters earlier — the same 'intelligent Agent' Newton invoked in his letters to Bentley reappears here, at the far end of the Opticks, applied now to the smallest particles of matter rather than to the arrangement of planets. Newton's science and his theology were never really separate books running in parallel; they were, on the evidence collected across this volume, chapters of the same single inquiry, however differently each one has come to be remembered.

It is a fitting note on which to end this pair of chapters devoted to wonder. Newton's certainties, examined earlier in this book, changed the shape of physics within his own lifetime. His questions took a good deal longer to bear fruit — decades, in the case of chemical affinity, and in a few cases, centuries. But he set them down anyway, with the same care he gave to his proofs, trusting that a well-posed question outlives the person who first asked it.

CHAPTER 10

The Mint

"About seven years ago Chaloner was a poor man, a Iapanner in cloaths thredbare ragged & daubed with colours & then got stamps made... & in a short time put on the habit of a Gentleman."

— Newton's case history of William Chaloner, Mint Papers, c. 1698

In 1696, at the age of fifty-three, Newton left Cambridge for London to become Warden of the Royal Mint — a post most of his predecessors had treated as a comfortable sinecure, collecting a salary while deputies did the actual work. Newton did not treat it that way. England's silver coinage was in crisis, debased for years by clipping and counterfeiting, and Newton threw himself into the Great Recoinage with the same total absorption he had once given to prisms and planets.

What followed was a second career few of his biographers know what to do with: Newton the detective. As Warden, and later as Master, he was responsible for prosecuting counterfeiters — a duty punishable, under a new 1696 law, by death, since coining was treated as an act of treason against the Crown. Newton went undercover in London's taverns and prisons, personally conducted more than a hundred interrogations, and built the cases, in his own hand, that sent dozens of counterfeiters to the gallows.

His greatest adversary was a resourceful, self-taught forger named William Chaloner, who spent years staying one step ahead of the Mint — at one point even accusing Newton's own office of corruption before Parliament, forcing Newton to defend his own department's integrity to a Parliamentary committee before he could resume the prosecution. Newton pursued him for nearly three years. This chapter draws from Newton's own surviving case notes: not the treatise-writer's voice, but a warden's, compiling evidence in prose as vivid as anything else he wrote.

The scale of the surviving archive is worth pausing on. Newton's Mint papers, held today at the National Archives at Kew, run to nearly two thousand separate documents — depositions, examinations, case summaries, correspondence with informants — almost all of it unpublished in his lifetime and largely unexamined by historians until relatively recently. For three decades, in other words, Newton was quietly compiling one of the more remarkable single-author criminal archives of the early modern period, alongside everything else he was doing as Master of the Mint and President of the Royal Society.

> *"About seven years ago Chaloner was a poor man, a Iapanner in cloaths thredbare ragged & daubed with colours & then got stamps made... left of his trade turned coyner & in a short time put on the habit of a Gentleman took a great house at Kinghtsbridge & bought some plate."*
>
> — 'Chaloner's Case': Newton's own history of Chaloner's crimes, Mint Papers, c. 1698

Newton, summarizing years of investigation into a prosecutable case, opens with something closer to a novelist's economy than a warden's report — a poor tradesman's sudden, suspicious transformation into a gentleman.

> *"Chaloner confessed to the Warden of the Mint that they were for casting of blanks & were*

> *thicker at one end than in other places to make room for the gate or gutter where the mettle was to run in."*
>
> — 'Chaloner's Case,' Mint Papers, c. 1698

Newton records, in the third person, his own interrogation of the suspect — a small, telling habit of a man for whom even personal involvement in a case was set down as evidence, precisely and without embellishment.

> *"He was originally taught to coyn by Coffee & learnt all his art of him & Grosvener & at that time... put in a stock with Coffee for French Pistols which were conveyed into France by a Gentleman to put into the Kings treasury & also bought by Blackford Coppinger Peck Holloway Iones Carter & many others."*
>
> — 'Chaloner's Case,' Mint Papers, c. 1698

Newton's case history reconstructs an entire criminal network, naming associate after associate — the same appetite for exhaustive, connected detail visible in his astronomical tables, turned here toward the underworld of London counterfeiting.

Chaloner was convicted and hanged at Tyburn in March 1699. In the days before his execution he wrote directly to Newton, begging for mercy in language of real desperation: in one surviving letter he wrote simply that he expected to be 'murderd' unless Newton intervened. These letters survive in the same archive as Newton's own case notes, an unintended correspondence between hunter and hunted. Nothing in Newton's surviving papers suggests he ever answered, or that the date of the execution moved him to any recorded reflection at all. The Warden who had spent three years building an airtight case seems, once it closed, to have simply moved on to the next one.

It is tempting to see this chapter of Newton's life as a curiosity, disconnected from the physicist and the philosopher. It is better understood as the same mind turned to a different kind of evidence: patient, exhaustive, unwilling to accept a guess where a fact could be established instead — applied here not to the orbit of a comet, but to the truth of a man's crimes. The precision that once measured the bending of a ray of light through a prism is, in these documents, measuring the exact thickness of a counterfeit coin blank, and drawing the same kind of unhurried, evidence-first conclusion from it.

It is also worth remembering what the Great Recoinage actually achieved beyond any single prosecution. Under Newton's supervision, England's entire silver currency was called in, melted down, and reminted to a reliable, trustworthy standard within a few short years — a logistical undertaking on a national scale, carried out by a Cambridge mathematician with no prior administrative experience, and judged by economic historians since to have been a genuine success. The Warden's pursuit of individual counterfeiters like Chaloner was, in the end, only the most vivid part of a far larger project: rebuilding public trust in the very idea of money.

There is a final irony worth noting. Chaloner's own defense, in his final months, rested partly on claiming special expertise — he had, after all, once offered the Mint his own advice on how to prevent counterfeiting, positioning himself as a reformer rather than a criminal. Newton's case notes methodically dismantle this claim, tracing Chaloner's supposed expertise back to the same criminal networks he claimed to be helping the Mint defeat. It is, in its way, a small mirror of the larger method traced throughout this book: a plausible-sounding story, examined carefully enough, giving way to what the actual evidence showed.

CHAPTER 11

Hidden Work

"What ever I can conceive without a contradiction, either is or may bee made by something that is."

— Of Natures obvious laws & processes in vegetation, c. 1672

For more than two centuries after his death, the world knew Newton almost exclusively as the author of the Principia and the Opticks — the physicist, the mathematician, the reformer of the Mint. It was not until a 1936 Sotheby's auction, when a hoard of his private manuscripts came onto the market and were bought largely by the economist John Maynard Keynes and the scholar Abraham Shalom Yahuda, that the wider world learned Newton had spent decades of his life on alchemy: reading, copying, and experimenting with a body of secretive, symbol-laden chemical writing going back centuries.

Most of what survives from this pursuit is, honestly, Newton transcribing other authors — the Polish alchemist Sendivogius, the shadowy figure known as Philalethes, and others — into his own notebooks, word for word. This chapter does not draw on that material, because it is not, in any meaningful sense, Newton's own words. It draws instead on one distinct manuscript, 'Of Natures obvious laws & processes in vegetation,' which scholars agree is Newton's own attempt to synthesize what he had read into an original theory — his own early try at what he might have called, had the phrase existed in his century, a theory of everything.

The manuscript dates from around 1672, placing it squarely in the same period as the prism letter that opens this book — the same young man, in the same years, working simultaneously on the optics that would make him famous and on a private, unpublished theory of matter that would not become public knowledge for two and a half centuries. Newton drew no hard boundary between the two projects. Both, in his own mind, were attempts to find the active, hidden principles beneath the visible surface of things — light and color in one case, growth and transformation in the other.

It is worth being honest about the scale of what Newton actually left behind in this area. Estimates place his surviving alchemical writings at well over a million words — more than he wrote on optics and gravity combined. For three centuries this fact sat almost entirely outside the public record, hidden not by any single act of concealment but by generations of biographers who preferred a tidier story: Newton the rationalist, the founder of modern physics, unburdened by anything that looked, to later eyes, like superstition. The manuscripts recovered in 1936 forced a more complicated and, in the end, more honest picture.

> *"What ever I can conceive without a contradiction, either is or may bee made by something that is: I can conceive all my owne powers (knowledge, activating matter etc.) without a contradiction."*
>
> — Of Natures obvious laws & processes in vegetation, c. 1672 (Dibner Collection MS. 1031B)

In a brief passage headed 'Of God,' Newton reasons — in language echoing the philosophical debates of his day about Descartes' proof of God's existence — from what can be coherently conceived to what might actually exist.

Historians no longer treat Newton's alchemy as an embarrassing footnote to be explained away. His interest was not superstition tacked onto real science; it was, in his own mind, part of the same investigation — a search for the active principles that make inert matter grow, change, and combine, pursued through whatever sources he could find, however strange those sources look from three hundred years distant. He kept almost none of it secret out of shame; alchemy itself, as a discipline, prized secrecy, and Newton simply followed its convention of writing for an audience of fellow initiates rather than the public.

It was John Maynard Keynes, reading through the recovered manuscripts in the 1930s, who offered the description of Newton that has stuck ever since: not the first of the modern rationalists, as the Enlightenment liked to remember him, but 'the last of the magicians' — a mind still reaching, by candlelight, for the same secrets the alchemists before him had spent centuries chasing. Keynes meant it as a corrective, not an insult. The Newton who could write with total clarity about the motion of the planets was, in his private notebooks, still willing to sift through obscure, allegorical texts on the chance that some grain of real knowledge about the workings of matter was buried inside them.

What this single manuscript shows, more clearly than the pages of transcription that surround it, is a mind unwilling to leave even its strangest and least provable pursuits entirely unexamined — reaching, in his own words this time, for the same first causes he would spend the rest of his life chasing through mathematics instead.

It is worth resisting, in closing, both of the easy stories this material invites. It would be too simple to say the alchemy was secretly the 'real' Newton, with the physics as some kind of public mask. It would be equally too simple to dismiss the alchemy as an embarrassing lapse in an otherwise flawless rational mind. The more honest reading, and the one this book has tried to hold to throughout, is that Newton pursued every question that genuinely interested him with the same total seriousness, whether or not the tools available to answer it were the ones that would later be called scientific.

There is also a practical postscript worth noting. Some of Newton's alchemical experimentation took place in a small furnace he kept and operated himself, at Cambridge and later in London, and there is credible speculation among some historians that his long exposure to mercury and other substances handled in the course of this work may have contributed to episodes of illness and volatile temperament recorded later in his life. Nothing in the surviving evidence proves this conclusively, and this book will not claim more certainty than exists. But it is a useful reminder that the alchemy was not merely theoretical reading — it involved real fire, real metal, and real physical risk, pursued by the same hands that ground the Opticks prisms and cast the Mint's silver.

CHAPTER 12

Comets and the Heavens

"To what end are Comets, and whence is it that Planets move all one and the same way in Orbs concentrick, while Comets move all manner of ways in Orbs very excentrick?"

— Opticks, Book III, Query 28

Before Newton, comets were widely treated as omens or anomalies — bright, unpredictable visitors that seemed to break whatever rules governed the orderly motion of the planets. The Principia's third book, 'The System of the World,' brought comets inside the same mathematics that explained everything else: the same law of gravity, Newton showed, that kept the earth in orbit around the sun also governed the long, eccentric paths of comets sweeping in from the outer reaches of the solar system and back out again.

This chapter draws from the Principia's account of the heavens as a single connected system, and from the Queries in which Newton, having proven what he could prove, allowed himself to wonder about the parts of the pattern he could not yet fully explain.

It is worth remembering the scale of what Newton was claiming. A comet, to the naked eye, looks nothing like a planet — a smeared, unpredictable streak of light rather than a steady point tracing a fixed circle. For most of recorded history, comets were treated as a wholly separate category of thing, closer to a weather event or a divine sign than to an orbiting body. Newton's friend Edmond Halley, working from the Principia's methods, would go on to show that a comet observed in 1682 was in fact the same body that had appeared in 1531 and 1607, returning on a long, regular, calculable orbit — the first confirmed prediction of a comet's return in history, and a direct application of the very mathematics collected in this chapter.

> *"By the propositions mathematically demonstrated in the former Books, in the third I derive from the celestial phenomena the forces of gravity with which bodies tend to the sun and several planets. Then from these forces, by other propositions which are also mathematical, I deduce the motions of the planets, the comets."*
>
> — Preface to the Principia (1687)

Newton describes, in his own summary of the book's structure, exactly how comets fit into the same chain of reasoning as everything else: not a special case requiring a separate explanation, but a direct consequence of the same force.

> *"To what end are Comets, and whence is it that Planets move all one and the same way in Orbs concentrick, while Comets move all manner of ways in Orbs very excentrick; and what hinders the fix'd Stars from falling upon one another?"*
>
> — Opticks, Book III, Query 28

Having explained the mathematics of how comets and planets move, Newton turns to a harder question — why the two categories of object are arranged so differently in the first place, and why the whole system of stars does not simply collapse under its own gravity.

> *"The same power whether natural or supernatural, which placed the Sun in the center of the orbs of the six primary Planets, placed Saturn in the center of the orbs of his*

> *five secondary Planets & Iupiter in the center of the orbs of his four secondary ones."*
>
> — First letter to Richard Bentley, 10 December 1692

The nested, repeated structure of the solar system — planets circling the sun, moons circling planets, all in the same tidy pattern — was, for Newton, as much a subject of wonder as anything he could see through a telescope.

That last question — what keeps the stars from falling on one another — would trouble Newton for the rest of his life, and he never resolved it to his own satisfaction; it appears again, worked at from a different angle, in his correspondence with Richard Bentley. It is one of the clearest examples in his whole body of work of a pattern this book has traced from its first chapter: mathematics carried as far as it could reliably go, and then, at the edge of what could be proven, an honest question left standing in its place.

It would be another two centuries before anyone could offer a fuller answer — modern cosmology's account of an expanding universe, with galaxies receding from one another fast enough to outrun their own gravity, was as unavailable to Newton as radio telescopes or space probes. What he had instead was the patience to notice the problem clearly, state it precisely, and leave it exactly where his evidence ran out, for whoever came next to pick up.

Halley's confirmed prediction, mentioned above, deserves one further word, since it says as much about Newton's method as about the comet itself. A prediction that comes true on a fixed, calculable schedule — as opposed to a description that merely fits what has already happened — is one of the sternest tests a scientific theory can face. The comet duly reappeared in 1758, sixteen years after Halley's own death, exactly where the Newtonian mathematics said it would. It remains one of the cleanest public vindications any theory in this book received, delivered decades after both men who had worked it out were gone.

It is worth remembering, too, how much of the Principia's third book is really an extended demonstration that the heavens follow no separate rules of their own. Aristotle's cosmology, still influential when Newton was a student, held that the celestial realm was made of a fundamentally different, incorruptible substance than the earth below — that the same laws simply could not apply to both. Newton's mathematics dissolved that distinction entirely. An apple falling in an orchard and a comet sweeping past the sun are, on his account, obeying the identical law, differing only in the numbers plugged into the same equation. Few single ideas in the history of science have collapsed so much inherited certainty at a stroke.

CHAPTER 13

The Unseen

"Does it not appear from Phaenomena that there is a Being incorporeal, living, intelligent, omnipresent, who in infinite Space, as it were in his Sensory, sees the things themselves intimately, and thoroughly perceives them?"

— Opticks, Book III, Query 28 (4th ed., 1730)

The longest and strangest of Newton's published Queries is also the one where his science and his theology come closest to merging into a single argument. Having spent the Opticks demonstrating, through exhaustive experiment, what could be proven about light and matter, Newton uses one of the final Queries to ask what those proofs might suggest about the maker of the system they describe — reaching for an image, borrowed from the emerging science of the human eye and nerve, to describe how God might perceive the universe.

This is delicate material, and Newton himself seems aware of it; the passage is phrased, like all the Queries, as a question rather than a claim. But the imagery is deliberate and carefully built, and it represents the furthest Newton allowed his published work to reach toward describing the nature of the divine.

The image Newton reaches for is drawn from the science of his own century's understanding of vision and the nerves — the idea, current among the physicians and natural philosophers he read, that sensation is not scattered throughout the body but gathered at some central place, a 'sensory,' where it is finally registered and known. Newton's move is to take that ordinary anatomical idea and scale it up to infinity: what if space itself is God's sensory, and everything that exists within it is, for God, as immediately present as the sensation of touch is to a human hand?

> *"Was the Eye contrived without Skill in Opticks, and the Ear without Knowledge of Sounds? How do the Motions of the Body follow from the Will, and whence is the Instinct in Animals?"*
>
> — Opticks, Book III, Query 28 (4th ed., 1730)

Newton opens with the same argument from design that runs through his letters to Bentley — the intricate, purposeful construction of a living creature's senses, offered as evidence against a purely mechanical universe.

> *"Is not the Sensory of Animals that place to which the sensitive Substance is present, and into which the sensible Species of Things are carried through the Nerves and Brain, that there they may be perceived by their immediate presence to that Substance?"*
>
> — Opticks, Book III, Query 28 (4th ed., 1730)

Newton first establishes the idea of a 'sensory' — a place where sensation actually registers — in ordinary animal perception, before extending the same image outward to its largest possible scale.

> *"And these things being rightly dispatch'd, does it not appear from Phaenomena that there is a Being incorporeal, living, intelligent, omnipresent, who in infinite Space, as it were in his Sensory, sees the things themselves intimately, and thoroughly perceives them, and comprehends them wholly by their immediate presence to himself."*
>
> — Opticks, Book III, Query 28 (4th ed., 1730)

The image completed: infinite space itself imagined as God's sensory organ, within which everything that exists is directly and immediately present to him — not perceived at a distance, as human senses perceive the world, but known completely and without mediation.

This passage — sometimes called Newton's 'sensorium' argument — drew sharp criticism in his own lifetime, most notably from Gottfried Leibniz, who thought it reduced God to something uncomfortably close to a being who needed organs to perceive the world at all. Newton never fully answered the objection to anyone's satisfaction, including, it seems, his own; he revised and softened the surrounding language across later editions without ever removing the core image.

The disagreement with Leibniz over this passage was not an isolated skirmish. It ran alongside, and fed into, their much larger and far bitterer dispute over who had first invented the calculus — a rivalry this book has deliberately kept at arm's length elsewhere, for reasons a later chapter explains. It is worth knowing, even so, that the two men's objections to each other were rarely confined to a single subject; a disagreement about the nature of God's perception and a disagreement about the priority of a mathematical discovery could, and did, become tangled up in the same long-running argument between two men who had, by this point, stopped extending each other much good faith.

Whatever its philosophical weaknesses, the passage remains a rare and valuable thing: a working scientist's own attempt, in his own words, to describe what an infinite, unmediated form of knowledge might be like — offered not as doctrine, in the manner of a theologian, but as one more Query, arising naturally, in his view, from everything else he had managed to prove.

It is worth placing this passage beside the Bentley letters examined earlier in this book. Both reach toward the same conclusion — an intelligent, omnipresent cause behind the visible universe — but by different roads. The Bentley letters argue from absence: from what Newton's own physics could not explain about the arrangement of the solar system. The sensorium passage argues from presence: from an image, borrowed from the science of perception, of what total and immediate knowledge might actually be like. Together they represent the two directions Newton's theology consistently ran — toward the limits of what could be proven, and toward the furthest reach of what could be imagined once those limits were honestly acknowledged.

It is worth noting, finally, that Newton chose to place this material not in a private letter or an unpublished manuscript, but inside the single most widely read scientific book he ever wrote, revised across four editions over more than a quarter century. Whatever caution he showed elsewhere about separating public certainty from private speculation, he judged this particular image worth the risk of publication — worth attaching his hard-won scientific reputation to a genuinely speculative claim about the nature of God's knowledge. Few passages in this book show more clearly how little distance, in Newton's own mind, actually separated his physics from his faith.

CHAPTER 14

The Nature of God

"God is a relative word & signifies much the same thing with Lord & King, but in a higher degree."

— Theological manuscript, Yahuda MS 15.5

Newton kept a secret his entire adult life that would have ended his career, and possibly endangered his liberty, had it become widely known: he did not believe in the Trinity. His private theological manuscripts — unpublished in his lifetime, dispersed at auction in 1936, and now held mainly at the National Library of Israel — show a man who had studied Scripture and early Church history with the same relentless thoroughness he brought to optics, and had concluded that the doctrine of three persons in one God was a corruption introduced centuries after Christ.

This is sensitive, doctrinally contested ground, and this chapter does not attempt to argue Newton's case or relitigate a fourth-century theological dispute. It offers instead a single, carefully chosen passage — Newton's own reflection on what the word 'God' itself means — general enough to stand on its own, without requiring agreement on the wider argument it was originally part of.

It is worth understanding what was actually at stake for Newton in keeping this material private. Denial of the Trinity was not merely an unfashionable opinion in late seventeenth-century England; it was, under the law of the time, heresy, and had been treated as a capital offense within living memory. As a Fellow of Trinity College, Cambridge, Newton would ordinarily have been required to take holy orders in the Church of England — a step that would have obliged him to affirm the very doctrine he privately rejected. He obtained, through influential friends, a rare royal dispensation excusing him from ordination, and never publicly explained why he needed one. The manuscripts recovered in 1936 are, in effect, the explanation he never gave in his own lifetime.

> *"God is a relative word & signifies much the same thing with Lord & King, but in a higher degree. As we say my Lord, our Lord, your Lord, the Lord of Lords, the supreme Lord... so we may say my God, our God, your God, the God of Gods, the supreme God."*
>
> — Theological manuscript, Yahuda MS 15.5, f. 98r

Newton reasons about the word 'God' the way a careful reader of any text reasons about a word used in many different relationships — asking what the term itself actually claims, before asking what else might be built on top of it.

Whatever a reader makes of the wider theological argument Newton was constructing, this single passage shows the same instinct visible everywhere else in his work: before reasoning about a thing, define your terms with total precision. He treated the word 'God' with the same discipline he had once applied to the word 'ray' at the opening of the Opticks — unwilling to build any larger claim on a foundation left vague.

The private, unpublished nature of this material is itself worth sitting with. Newton wrote hundreds of pages on subjects he never intended the public to see, at real personal risk, apparently because he could not stop pursuing a question once he had begun — the same compulsion that gave the world the calculus and the law of gravity, turned here toward a subject he judged too dangerous to publish, and too important to leave unexamined.

It is worth adding, too, what this material does not show. It does not show a man losing faith, or drifting toward disbelief in his later years, as some popular accounts have suggested. If anything, the opposite is true: the manuscripts reveal a Newton whose religious convictions were even more thoroughly worked out, and more personally consequential, than his careful public silence on the subject ever let on.

It is worth adding a final observation about the scale of Newton's biblical scholarship, since the single quotation collected in this chapter can only hint at it. He compiled extensive chronological tables cross-referencing scriptural genealogies against secular history, studied biblical Hebrew and the manuscript traditions of the New Testament with the same rigor a modern textual critic would apply, and produced, among much else, a lengthy manuscript examining two specific biblical passages he believed had been corrupted centuries after their original composition specifically to support Trinitarian doctrine. Whatever a reader makes of his conclusions, the sheer volume and seriousness of the underlying scholarship rules out any suggestion that this was casual or half-formed thinking.

CHAPTER 15

Patience

"To explain all nature is too difficult a task for any one man or even for any one age."

— Manuscript note related to the Opticks Queries

Newton worked slowly, by his own account, and distrusted quick answers. The Principia took years to write once he finally sat down to it, after decades of the underlying ideas sitting half-formed in his notebooks. The Opticks was delayed from publication for years beyond its completion, in part because Newton dreaded the disputes that had followed his first paper on light. This chapter gathers the moments where Newton reflected, in his own words, on the value of going slowly — of doing a smaller thing certainly, rather than a larger thing on guesswork.

It is a useful corrective to the popular image of Newton as a man struck by sudden inspiration — the falling apple, the flash of insight. The surviving record shows something closer to the opposite: ideas held for years, turned over, tested, set aside, and returned to, sometimes across a decade or more, before Newton judged them ready for other people to see. The apple story itself, to the extent it happened at all, describes only the first spark of an idea that then took the better part of twenty years of quiet, unglamorous work before it became the law of gravity presented in the Principia.

> *"To explain all nature is too difficult a task for any one man or even for any one age. 'Tis much better to do a little with certainty & leave the rest for others that come after, than to explain*

all things by conjecture without making sure of any thing."

— Manuscript note related to the Opticks Queries

The clearest surviving statement of Newton's working philosophy, set down privately rather than for publication — advice, in effect, to himself.

"Part of the ensuing Discourse about Light was written at the Desire of some Gentlemen of the Royal-Society, in the Year 1675, and then sent to their Secretary, and read at their Meetings, and the rest was added about twelve Years after to complete the Theory."

— Advertisement I, Opticks (1704)

In his own preface, Newton admits the plain arithmetic of how long the book actually took: a first portion written in 1675, the remainder added roughly twelve years later, before the whole thing was finally allowed to go to print.

"To avoid being engaged in Disputes about these Matters, I have hitherto delayed the printing, and should still have delayed it, had not the Importunity of Friends prevailed upon me."

— Advertisement I, Opticks (1704)

Newton names his own reluctance directly: it was not the work that took so long, but his dread of the arguments that publication would inevitably reopen.

"If any other Papers writ on this Subject are got out of my Hands they are imperfect, and were perhaps written before I had tried all the Experiments here set down, and fully satisfied my self about the Laws of Refractions and Composition of Colours."

— Advertisement I, Opticks (1704)

Newton worries, in print, about earlier drafts of his own work circulating without his permission — a reminder that even the finished, published Opticks represents a version he considered complete only after many earlier, discarded attempts.

It is worth remembering, reading these admissions, that they come from the same man capable of astonishing bursts of speed — the plague years at Woolsthorpe produced, in a matter of months, insights that took him decades to fully work out and publish. Newton's patience was not slowness of thought. It was a refusal to let even a fast, confident insight go to print before it had been tested as thoroughly as he knew how to test it, however many years that took.

There is a cost to this kind of patience, and Newton paid it more than once. Gottfried Leibniz published his own version of the calculus years before Newton's Method of Fluxions finally appeared in print, and the resulting priority dispute — real, bitter, and never fully resolved in either man's lifetime — was in no small part a consequence of exactly the delay described in this chapter. Newton's caution protected the quality of his published work. It did not always protect his claim to have been first.

It is worth asking, in fairness, whether the trade was a bad one. The Opticks that finally appeared in 1704 is a far more finished, far more defensible book than anything Newton could have published in 1672, in the immediate aftermath of the Oldenburg letter and the dispute with Hooke that followed it. Patience cost him priority on more than one occasion. It also meant that when his major works finally did appear, they were built to withstand exactly the kind of scrutiny a hurried publication could not have survived.

It is worth adding a further, more mundane explanation alongside the psychological one this chapter has emphasized: much of Newton's delay was simply logistical. Cambridge in this period had no reliable scientific printing industry of its own, and coordinating publication from a distance, revising proofs by post, and settling on illustrations for a technical book was a genuinely slow process by the standards of any century, let alone the seventeenth. Some of the patience credited to Newton's temperament in this chapter was, in fair measure, also patience forced on him by the ordinary friction of getting a difficult book into print at all.

CHAPTER 16

Scripture and Time

"I admit no Chronology of things done in Europe, above eighty years before Cadmus brought letters into Europe."

— The Chronology of Ancient Kingdoms Amended (1728)

Newton spent decades, on and off across his life, trying to establish a more accurate chronology of the ancient world — working from astronomical records, king lists, and his own reconstructions of when various ancient events must have occurred, in order to correct what he believed were centuries of accumulated error in the accepted history of Egypt, Greece, Assyria, and Persia. He was still revising the resulting manuscript, The Chronology of Ancient Kingdoms Amended, at the time of his death in 1727; it was published the following year.

The book's specific conclusions have not held up well — Newton dated events using methods and assumptions modern historians no longer accept, and some of his sources treated as historical fact are now understood as legend. He argued, among other things, that Solomon's Temple in Jerusalem was the earliest true temple ever built, with all other ancient temples copied from it, and he tried to fix precise dates for the Argonauts' voyage and the Trojan War using calculations based on the position of the stars described in later, secondhand accounts. What is worth preserving from the project is not its results but its method: the same skepticism toward unverified claims that drove his physics, turned here toward the deep past.

Newton's chief tool for this project was astronomy rather than textual criticism. Ancient writers occasionally recorded the position of a constellation at the time of some historical event — a coronation, a battle, the founding of a city — and Newton, knowing how the positions of the stars slowly shift over centuries due to a phenomenon called precession, tried to work backward from those descriptions to calculate an exact date. It was an ingenious method in principle. In practice, the ancient descriptions he relied on were often too vague, or too poetic, to support the precision he tried to extract from them — and this, more than any single error, is why the finished work convinced few of the historians who read it even in his own century.

> *"Chronologers have split Apis Epaphus or Epopeus into two Kings, whom they call Apis and Epopeus, and between them have inserted eleven or twelve feigned names of Kings who did nothing, and thereby they have made its Founder Aegialeus, three hundred years older than his brother Phoroneus."*
>
> — The Chronology of Ancient Kingdoms Amended (1728)

Newton catalogues, with evident irritation, the kind of error he believed had crept into accepted ancient chronologies — invented kings, padded reigns, and lists stretched out far beyond what any real evidence supported.

> *"Some have made the Kings of Germany as old as the Flood: and yet before the use of letters, the names and actions of men could scarce be remembred above eighty or an hundred years after their deaths: and therefore I admit no Chronology of things done in Europe, above eighty years before Cadmus brought letters into Europe."*
>
> — The Chronology of Ancient Kingdoms Amended (1728)

Newton's governing principle, stated plainly: without written records, oral memory of real events does not reliably survive more than a lifetime or two — so any chronology claiming to reach further back than that, in a society without writing, should be treated with suspicion.

It is a strange thing to watch the author of the Principia apply his exacting standards of evidence to the record of the ancient world and arrive, by his own reckoning, at conclusions modern scholarship has since set aside. But the impulse behind the project deserves more respect than the project's results. Newton distrusted a king list padded with invented names for the same reason he distrusted a physical theory built on invented mechanisms: both dressed up guesswork as established fact, and both, in his view, deserved to be corrected rather than politely left alone.

It is worth adding, honestly, that this is the clearest case in this book of Newton's method outrunning his evidence. The instinct — do not trust a claimed date without checking what actually supports it — was sound. The available evidence from the ancient world, three thousand years removed and filtered through centuries of retelling, simply could not bear the mathematical precision Newton tried to build on top of it. Even the most rigorous method produces only as much certainty as its underlying evidence allows, and nowhere in his body of work is that limit more visible than here.

There is, even so, something worth admiring in the project's sheer persistence. Newton returned to this manuscript across five decades, revising and re-revising a work he must have known, at some level, would never satisfy the scholars who actually specialized in ancient history. He kept at it anyway — not out of stubbornness alone, one suspects, but because the same restless unwillingness to leave a question unexamined that produced the calculus and the law of gravity did not switch itself off simply because the subject had changed from physics to the deep human past.

It is worth noting, finally, the peculiar afterlife this project had once it was finally published. The Chronology sparked immediate scholarly controversy across Europe, with French and German historians publishing detailed rebuttals within a few years of its 1728 appearance — a lively international dispute conducted almost entirely without Newton himself present to defend it, since he had died the year before publication. It stands as one of the few arguments in this book that Newton set in motion but never lived to see through to any kind of resolution, contested or otherwise.

CHAPTER 17
Numbers

"Mathematical quantities I here consider not as consisting of indivisibles... but as described by a continued motion."

— Method of Fluxions (written 1671, published 1736)

Newton developed the mathematics now called calculus during the plague years at Woolsthorpe, working out how to handle quantities that change continuously — the speed of a falling body at a single instant, the area under a curving line — years before he had any practical use for the technique. He called it the method of fluxions, and characteristically, he did not rush to publish it; the full treatise did not appear in print until 1736, nine years after his death, though the mathematics had been circulating in fragments and letters for decades before that.

This chapter draws on Newton's own account of the foundations of his method — a rare glimpse of him reasoning about mathematics itself, rather than using mathematics to reason about the physical world.

It is worth being clear about what fluxions actually let Newton do, since the word itself has fallen out of use. A fluxion, in his terminology, is a rate of change — how fast a quantity is growing or shrinking at a single instant, rather than averaged over some longer stretch of time. Before this method existed, there was no reliable way to answer a question like: exactly how fast is a falling object moving right now, at this precise moment, rather than on average across the whole fall? Newton's calculus made that kind of question answerable, and it turned out to be exactly the tool needed to describe motion under a continuously acting force like gravity — which is part of why the mathematics and the physics in this book were never really separate projects.

> *"Mathematical quantities I here consider not as consisting of indivisibles, either parts least possible or infinitely small, but as described by a continued motion... each time it can conveniently so be done, it is preferable to express them by finite lines visible to the eye rather than by infinitely small ones."*
>
> — Method of Fluxions (written 1671, published 1736)

Newton explains his own discomfort with a foundation built on infinitely small quantities — a discomfort that would eventually become one of the sharpest philosophical objections raised against calculus itself, by critics who felt neither Newton's nor Leibniz's version of the new mathematics had been placed on fully solid ground.

> *"For fluxions are finite quantities but moments here are infinitely little."*
>
> — Principia, second edition, 1713

Even decades later, revising the Principia, Newton is still carefully distinguishing between two related but different ideas in his own system — the finite, measurable rate of change, and the infinitely small increment used to define it — anxious, as ever, not to let a useful shorthand be mistaken for a claim he could not defend.

The objection Newton anticipated in his own foundational writing — that calculus rested on quantities too small to be properly defined — would not be fully answered until the nineteenth century, when mathematicians developed the rigorous concept of a limit. Newton did not have that tool. What he had instead was the same instinct visible throughout this book: a preference, wherever possible, for what could be pictured and checked over what could only be asserted, applied here to the strange, powerful, and at the time genuinely unsettled new mathematics he had built almost entirely on his own.

The irony is that Newton's own caution about publishing this work is part of why the credit for it became so contested. Gottfried Leibniz, working independently in Germany, developed his own version of the same calculus and published it years before Newton's Method of Fluxions finally appeared in print. Both men, by every serious modern historian's judgment, arrived at the idea on their own. But the bitterness of the priority dispute that followed — a fight this book has chosen largely to leave outside its pages — was fed in no small part by exactly the years of delay described here.

It is worth remembering, finally, what this mathematics was actually for in Newton's own hands. He did not develop fluxions as an abstract exercise for its own sake. He developed it because the physics he was trying to describe — falling bodies, orbiting planets, the changing speed of anything moving under a continuously acting force — could not be properly captured by the geometry and algebra already available to him. The calculus and the Principia grew up together, each one demanding refinements in the other, and it is difficult to say, in the end, which was really in service of which.

It is worth noting, too, a curious feature of the Principia's own published mathematics: Newton largely avoided using his own calculus notation in the finished book, presenting almost all of his proofs instead in the older, more familiar language of classical geometry. Historians have long debated why — whether it was a deliberate choice to make the work more persuasive to readers trained in Euclid, or evidence that Newton had originally worked out many of the results using geometric methods before calculus entered the picture at all. Either way, the book that did more than any other to demonstrate the power of his new mathematics is, on its own pages, strikingly reluctant to show that mathematics directly.

CHAPTER 18

Instruments

"An accompt of a new catadioptrical telescope invented by Mr. Newton, Fellow of the R. Society, and Professor of the Mathematiques in the University of Cambridge."

— Philosophical Transactions of the Royal Society, 1672

Newton's fame at the Royal Society began not with his theory of light but with an object: a reflecting telescope, barely six inches long, that he had designed, cast, and ground with his own hands at Cambridge. Where earlier telescopes used a lens to gather light — and suffered from a blurring, colored fringe around every image, a defect Newton's own optical experiments would later explain — his small instrument used a curved mirror instead, avoiding the problem entirely while achieving a magnification that astonished the Fellows who examined it in London.

The problem Newton was solving was well known to every telescope-maker of his generation. A simple glass lens does not bend every color of light by exactly the same amount — a fact this book has already traced back to Newton's own prism experiments — and the result, in a telescope, is a faint rainbow-colored halo smeared around every bright object, blurring fine detail no matter how carefully the lens was ground. Newton's insight, developed before he had even fully published the theory of color that explained the problem, was to replace the main lens with a curved mirror instead. A mirror reflects every color at exactly the same angle, so the defect disappears entirely.

He built the instrument almost entirely by hand: grinding and polishing the mirror himself from an alloy of his own devising, built up in a small home workshop rather than commissioned from a professional instrument-maker, as was then the custom for gentlemen of his standing. This chapter draws on the record of that invention: a rare case of Newton being celebrated first for something he had built, before he was known for anything he had proven.

> *"An accompt of a new catadioptrical telescope invented by Mr. Newton, Fellow of the R. Society, and Professor of the Mathematiques in the University of Cambridge."*
>
> — Title of Newton's submission, Philosophical Transactions of the Royal Society, 1672

The formal announcement of the instrument that won Newton his Fellowship — a rare case where the record of the achievement is, in effect, its own headline.

> *"I have here sent you a further Description of the Instrument for seeing by Reflection, together with an accompt of some further advantages, and defects of it, which possibly you have not been informed of."*
>
> — Letter to Henry Oldenburg, 1672

As with the prism experiments, Newton's account of the telescope is candid about the instrument's limitations as well as its strengths — a habit of naming a device's defects alongside its advantages that runs through nearly everything he built.

Newton's telescope was demonstrated before the Royal Society at the end of 1671 and caused, by every contemporary account, a genuine sensation: a tube shorter than a man's forearm outperforming refracting telescopes many times its length. It was this small, physical accomplishment — something the Fellows could hold, examine, and look through with their own eyes — that opened the door for the far more disruptive theory of light and color Newton sent them only weeks later. He had, in effect, earned the right to be believed before he ever needed to be.

The design Newton pioneered proved durable well beyond his own lifetime. Reflecting telescopes built on the same basic principle — a curved mirror in place of a primary lens — remain the standard design for large observatory and space telescopes to this day, scaled up from Newton's six-inch tabletop instrument to mirrors many meters across. Few of Newton's inventions, as opposed to his discoveries, are still in everyday professional use three and a half centuries later; this modest object, built by an obscure Cambridge mathematician in his own rooms, is one of them.

There is a pleasing symmetry in how this chapter connects to the second chapter of this book. Newton built the telescope to solve a physical problem — the colored fringe blurring every image — years before he had fully articulated, in the Oldenburg letter, exactly why that fringe appeared in the first place. The instrument came first; the explanation for why it worked came only afterward, once the prism experiments had run their course. Invention and theory, in this one case, arrived in the opposite order from how the story is usually told.

It is worth remembering, too, the sheer physical labor involved in producing the instrument at all. Grinding and polishing a mirror to the precision needed for useful magnification was skilled, repetitive manual work, typically the province of professional instrument-makers who guarded their techniques closely. Newton learned enough of the craft on his own — reportedly experimenting with his own alloy recipes to find a metal that would take a brighter polish than the speculum metal then in common use — to produce a working instrument without any formal training in the trade. It is a small, concrete example of a pattern visible everywhere else in this book: faced with a problem no existing tool could solve, Newton was rarely content to wait for someone else to build the tool for him.

CHAPTER 19

Teaching

"Since Sir Isaac's Lectiones Opticae, which he publickly read in the University of Cambridge in the Years 1669, 1670, and 1671, are lately printed, it has been thought proper to make at the bottom of the Pages several Citations from thence."

— Advertisement to the Fourth Edition of the Opticks

In 1669, at twenty-six, Newton was appointed Lucasian Professor of Mathematics at Cambridge — a chair he would hold for over three decades, though he was, by every surviving account, a strange and largely unsuccessful lecturer. His biographer and assistant Humphrey Newton (no relation) recorded that Newton sometimes lectured to an empty room, and left again within half an hour when no students appeared, rather than wait. The lectures he did give, however, were not wasted; written up as the Lectiones Opticae, they became an early working draft of the ideas that would eventually mature into the Opticks itself.

It is worth trying to picture the scene honestly. The Lucasian chair required its holder to deliver a set number of lectures each year and to deposit written copies in the university library — a bureaucratic obligation Newton fulfilled to the letter, even when almost no one showed up to hear them delivered. Cambridge in this period had few students with the mathematical background to follow Newton's material, and fewer still with the patience for his exacting, proof-by-proof style. He seems to have kept lecturing anyway, for over a decade, less because it was rewarding than because it was required — and, one suspects, because writing the lectures forced him to organize thinking he would otherwise have kept scattered across private notebooks.

This chapter is necessarily modest — the lectures are technical, geometrical, and written for an audience already expected to know the subject — but they capture something worth preserving: Newton in the act of teaching, however imperfectly attended, working out in front of students the same ideas he would later publish for the world.

> *"Since Sir Isaac's Lectiones Opticae, which he publickly read in the University of Cambridge in the Years 1669, 1670, and 1671, are lately printed, it has been thought proper to make at the bottom of the Pages several Citations from thence, where may be found the Demonstrations, which the Author omitted in these Opticks."*
>
> — Advertisement to the Fourth Edition of the Opticks (1730)

Newton's publisher, preparing the final edition of the Opticks after Newton's death, notes that the earlier lectures contain fuller proofs than the published book itself — Newton had, characteristically, trimmed the classroom version down for a wider readership, keeping the harder mathematics in reserve.

> *"The Rays of Light, in my Judgment, are not Modifications of any Homogeneal medium, but distinct Bodies, or at least distinct powers or Modifications of powers by which they are able to move & agitate the Aether."*
>
> — Lectiones Opticae, Cambridge lectures, 1670-72

In front of whatever students had gathered, Newton was already teaching, years before the Opticks reached print, the same central claim about the nature of light that would eventually make his public reputation — the classroom, in this case, running years ahead of the published book.

It is a small irony worth sitting with: the man whose Cambridge lectures went largely unheard by students in his own lifetime left behind, in the record of those same lectures, some of the clearest surviving evidence of how his greatest published work was actually built — proposition by proposition, in front of a room, whether or not the room happened to be full.

Newton's most famous student, in the end, was probably not any of the young men who sat, or failed to sit, in his Cambridge lecture hall. It was the wider republic of readers, across Europe and the centuries since, who encountered the same ideas fully worked out in the Opticks and the Principia — an audience Newton could not see and would never meet, but one that turned out, in the end, to be listening far more closely than the room in front of him ever had.

One student, at least, made a lasting difference simply by being present: Newton's assistant Humphrey Newton, who copied out manuscripts, prepared diagrams, and lived alongside him at Cambridge for several years during the period the Principia was written. It is largely from Humphrey's own later recollections, gathered by early biographers, that historians know anything at all about Newton's daily habits — his long hours, his frequent forgetting of meals, his rare laughter. An unsuccessful lecturer, it turns out, could still leave behind an attentive witness, simply by sharing a house with one.

It is worth adding, finally, that the Lucasian chair itself would go on to become one of the most celebrated academic positions in the world, held in later centuries by figures such as Charles Babbage, Paul Dirac, and Stephen Hawking — a lineage of occupants routinely invoked, in popular accounts, as evidence of the chair's prestige. Whatever that later prestige owes to Newton's own three decades in the post, it is worth remembering it was earned largely on the strength of what he published afterward, rather than on the attendance figures of the lectures he actually delivered while holding it.

CHAPTER 20
Explaining

"Newton, Cambridge, answering a question of Mr. Pepys, concerning the odds upon certain given chances at dice."

— Correspondence with Samuel Pepys, November–December 1693

In November 1693, Samuel Pepys — best known today as the diarist, though Newton would have known him as a former Secretary of the Admiralty and a past President of the Royal Society — wrote to Newton with a question he had been asked by a schoolmaster friend, John Smith, about a wager involving dice. Which was more likely: rolling at least one six with six dice, at least two sixes with twelve dice, or at least three sixes with eighteen dice?

Pepys, and Smith, guessed that more dice meant better odds. Newton, in an exchange of letters running into December, showed them they were wrong — and did it with a patience worth noticing, since Pepys pressed him more than once for a fuller explanation before he was satisfied.

The exchange is worth placing in context. Newton at this point was fifty years old, already the author of the Principia, and one of the most famous natural philosophers in Europe. Pepys was not a mathematician and had no particular claim on Newton's time beyond an existing acquaintance through the Royal Society. That Newton nonetheless worked through the problem carefully, more than once, for a correspondent outside his own field, says something about how he handled a question once someone had actually put it to him directly — a marked contrast to the guarded, sometimes prickly figure visible in his disputes with rival natural philosophers.

> *"Sir, Some time after Mr Millington had delivered your message, he pressed me to see you the next time I went to London."*
>
> — Letter to Samuel Pepys, 1693

Even in a short, informal note to a friend outside the world of natural philosophy, Newton's opening is careful and precise — an old habit, applied here to ordinary correspondence rather than to a proof.

> *"It being in my judgment, according to the doctrine of chances, that A has the best of the laying, and (3) is worse for B than either (1) or (2)."*
>
> — Letter to Samuel Pepys, 26 November 1693

Newton delivers his verdict plainly — the wager most people's intuition favors turns out to be the worst bet of the three — before walking Pepys through the reasoning that gets him there.

Newton reached the correct conclusion — the six-dice wager was in fact the safest bet — though his own stated reasoning, historians have since shown, contained a subtle logical slip that would not be caught and corrected for another three centuries. It is a fittingly human note to end on: even here, patiently walking a friend through a problem of dice and odds, Newton was neither infallible nor uninterested in getting the small things right — he simply kept working at the explanation until his correspondent had what he needed.

It is worth noting how different this correspondence feels from nearly everything else collected in this book. There is no priority to defend, no rival to answer, no theology at stake — only a friend's question, answered as clearly as Newton knew how, across several patient letters, about a problem whose only real consequence was a schoolmaster's wager. It is a small, welcome glimpse of a side of Newton not often on display: generous with his time, unguarded, and simply glad, it seems, to have a puzzle worth solving.

It is worth remembering, too, who Pepys was to Newton at this point: not a subordinate seeking instruction, but a former president of the very Royal Society Newton belonged to, and a man whose own diaries — though Newton could not have known they would one day be read by millions — already showed the same appetite for exact, curious observation of ordinary life that Newton brought to a prism or a planet. Two very different kinds of careful mind, meeting briefly over a question about dice, each recognizable to the other as a fellow close observer of the world.

It is worth adding a final, slightly wry note. The dice problem Newton solved for Pepys is now a standard textbook example in the teaching of probability theory, still assigned to students learning the subject more than three centuries later, usually without any mention of the correspondence's original, entirely informal context. A schoolmaster's idle wager, relayed through a retired naval administrator to a busy Cambridge professor, has quietly outlived every party involved to become part of the basic furniture of a field of mathematics that barely existed, in any rigorous form, when the letters were written.

CHAPTER 21

A Life's Work

"The Chronology of Ancient Kingdoms amended, was writ by the Author many years since; yet he lately revis'd it, and was actually preparing it for the Press at the time of his death."

— Publisher's advertisement, The Chronology of Ancient Kingdoms Amended (1728)

Newton died on the twentieth of March, 1727, at the age of eighty-four, having served as Master of the Mint and President of the Royal Society almost to the end. He was given a state funeral and buried in Westminster Abbey — the first scientist in England to be so honored. A great deal has been written, much of it unreliable, about what he is supposed to have said or felt in his final years. This book has tried, throughout, to hold that material to the same standard Newton held his own work: nothing offered as his words without a document to show for it.

What can be shown, plainly, is what he was still doing. In his final months, by his own publisher's account, Newton was not resting on the achievements already decades behind him. He was still at his desk, still revising, still trying to get one more piece of work — the long, difficult chronology of the ancient world — into a state he judged fit to publish.

It is worth remembering the full span this represents. Sixty years separate the darkened room at Woolsthorpe where a young man first turned a prism toward a beam of sunlight, and the desk in London where an old man made his final corrections to a chronology of ancient kingdoms he would not live to see printed. In between lay the

Opticks, the Principia, three decades at the Mint, a quarter century as President of the Royal Society, and the private theological and alchemical writings this book has drawn on only carefully and in part. Very few lives, in any era, cover that much ground without losing their thread. Newton's, on the evidence gathered across this whole book, never really did.

The funeral itself, by every surviving account, was a genuinely extraordinary public event. Voltaire, then visiting England, wrote with evident astonishment that he had seen a mathematician buried like a king who had been good to his subjects — a comparison that would have startled Newton's own Cambridge contemporaries, who for decades had known him mainly as an eccentric, reclusive professor. Six peers of the realm carried his pall. It is worth setting that public grandeur beside the private, unglamorous fact this chapter has tried to establish: that in his final months, the man being buried with such ceremony was, by all available evidence, simply still at his desk.

It is worth adding, finally, a brief accounting of what actually happened to Newton's papers after his death, since it explains why so much of this book's material has such an unusual, delayed publication history of its own. The bulk of his private manuscripts — theological, alchemical, and mathematical alike — passed down through family connections rather than into any institutional archive, and much of it sat largely unexamined in private hands for over two centuries. It was not scattered by accident so much as simply never made public: no single decision buried it, but no single decision brought it fully to light either, until the 1936 Sotheby's sale finally forced much of it into scholarly view. The Newton this book has tried to present in his own words was, in a very real sense, only fully assembled in the twentieth century, two hundred years after his death.

> *"The Chronology of Ancient Kingdoms amended, was writ by the Author many years since; yet he lately revis'd it, and was actually preparing it for the Press at the time of his death."*
>
> — Publisher's advertisement, The Chronology of Ancient Kingdoms Amended (1728)

The last confirmed, documented fact of Newton's working life: not a deathbed epigram, but an old man, still editing.

> *"To explain all nature is too difficult a task for any one man or even for any one age. 'Tis much better to do a little with certainty & leave the rest for others that come after, than to explain all things by conjecture without making sure of any thing."*
>
> — Manuscript note related to the Opticks Queries

It seems fitting to close this account of Newton's working life where the book itself began — with the same private note on method, written decades earlier, that Newton seems to have followed for the whole of the life recorded in these pages.

It would be easy, and is tempting, to close a book like this on a grander note — a final, resonant sentence pulled from an old man looking back on an extraordinary life. The most famous candidate for that sentence, a line about a boy playing on a seashore before a great, undiscovered ocean of truth, is not reliably Newton's own words at all; it surfaced in print more than a century after his death, attributed secondhand to a man who was, by the historical record, not even in England when he is supposed to have heard it. This book has chosen, throughout, to leave that sentence out.

What remains instead is smaller and, in its way, better evidence of the man: sixty years of a mind that would not stop checking its own work, right up to the very last document that survives. Newton spent his life refusing to accept an answer until it had been tested, and he seems to have died the same way — not finished, but still revising, still unwilling to send the last page to the printer until he was as certain of it as sixty years of practice had taught him to be.

That, more than any single quotable line, is the portrait this book has tried to leave behind: not a man who had all the answers, but one who could not stop asking better questions — of a prism, of a comet, of a coiner's confession, of a friend's dice game, and, in the end, of himself.

CHAPTER 22

Family

"Threatning my father and mother Smith to burne them and the house over them."

— Fitzwilliam Notebook, list of sins, c. 1662

Isaac Newton was born on Christmas Day 1642, three months after his father — also named Isaac, a yeoman farmer who could not sign his own name — had died. When Newton was three, his mother, Hannah, remarried a wealthy older clergyman named Barnabas Smith and moved to his rectory a mile away, leaving the young Newton behind at Woolsthorpe Manor to be raised by his grandmother. Smith did not want another man's child in his household, and the arrangement, negotiated between the adults, was final: for the next seven years, until Smith's death, Newton grew up apart from his mother, a short walk away, without her.

Newton left almost no direct commentary on how this felt. What survives instead is a single, extraordinary document: a notebook, kept around 1662 when he was nineteen and newly arrived at Cambridge, in which he set down a list of forty-nine sins committed 'before and after Whitsunday' of that year, written in a simple substitution cipher he had devised himself, evidently for privacy rather than concealment from any particular reader. Rediscovered only in the twentieth century, it is the closest thing that survives to Newton examining his own childhood in his own hand.

This chapter draws on that list, alongside a handful of other surviving fragments touching on Newton's early family life, and offers no interpretation Newton did not himself commit to paper. Where the record is genuinely silent, this book stays silent with it.

> *"Threatning my father and mother Smith to burne them and the house over them."*
> — Fitzwilliam Notebook, list of sins, c. 1662

Among forty-eight other, mostly minor confessions — eating an apple in church, making pies on a Sunday, stealing cherry stones from a schoolfellow — this single line stands apart. Newton records it in the same flat, numbered sequence as the rest, offering no explanation and no excuse. This book will follow his own restraint and offer none either.

> *"Punching my sister."*
> — Fitzwilliam Notebook, list of sins, c. 1662

A smaller, more ordinary entry from the same list, referring to one of the three younger half-siblings Hannah and Barnabas Smith had together — children Newton grew up alongside once his mother and stepfather's household eventually merged with his own after Smith's death in 1653.

> *"Refusing to go to the close at my mothers command."*
> — Fitzwilliam Notebook, list of sins, c. 1662

A brief note of ordinary childhood defiance, notable mainly for confirming that Hannah did return to active involvement in her son's upbringing once Smith died and she came back to Woolsthorpe — the relationship between them, whatever its earlier strain, evidently continuing into Newton's later boyhood.

> *"Setting my heart on money learning pleasure more than Thee."*

— Fitzwilliam Notebook, list of sins, c. 1662

As the list continues, the entries shift from specific childhood incidents to broader, recurring failings — less an inventory of events than a private accounting of character, the same self-examining habit of mind visible everywhere else in this book, turned here on himself at nineteen.

Barnabas Smith died in 1653, when Newton was ten, and Hannah returned to Woolsthorpe with her three younger children, reuniting the household Newton had been separated from for the better part of a decade. He was sent, not long after, to grammar school in nearby Grantham, lodging with a local apothecary's family rather than living at home — another departure, though a more ordinary one for a boy of his class and promise. He excelled academically once there, by every surviving account, though early biographers also recorded him as solitary and more interested in building mechanical toys — sundials, a model windmill, a water clock — than in the company of other boys.

It would be easy, and is tempting, to draw a straight line from these fragments to the adult Newton's well-documented difficulty with sustained personal attachment — his lifelong bachelorhood, his notoriously prickly relationships with rivals, his rare recorded expressions of warmth even toward friends and colleagues. This book has tried, throughout, to hold itself to the discipline Newton himself prized: no claim beyond what the evidence actually supports. The sin list shows a boy who once, in a moment of anger, wished harm on the mother and stepfather who had left him behind. It does not show us why, and it does not entitle a reader three and a half centuries later to decide what that anger explains about the man he became.

What can be said with more confidence is this: the same unflinching, exact self-examination Newton later brought to a prism, a planetary orbit, and a coiner's confession, he first turned on himself, at nineteen, in a private notebook no one else was meant to read. Before he was famous for holding the physical world to an exacting standard of evidence, he had already been holding himself to one.

CHAPTER 23
Rivalry

"Second inventors have no right. Whether Mr Leibniz found the Method by himself or not is not the Question."

— Newton's account of the Commercium Epistolicum, 1715

This book has, until now, deliberately kept one of the largest disputes of Newton's life at arm's length, mentioning it only in passing — his decades-long argument with the German philosopher Gottfried Wilhelm Leibniz over who had first invented the calculus. It has stayed at arm's length because the episode reflects, by the judgment of nearly every serious historian since, badly on Newton's character, and this book has not wanted to build a chapter that reads as either an attack on him or an excuse for him. But a book built on Newton's own words cannot avoid entirely what he actually wrote and did during the years this dispute consumed, and a full account of the man requires looking at it directly, however uncomfortable the material.

The bare facts, on which modern scholarship is largely settled, are these: Newton developed his method of fluxions in the mid-1660s, years before Leibniz began his own independent work on the same underlying mathematics in the 1670s. Leibniz, however, published first, in 1684, while Newton's own treatise did not reach print until 1736 — nine years after Newton's death. For decades the two men and their respective circles of supporters accused each other, with escalating bitterness, of plagiarism, culminating in a formal 1712 investigation by the Royal Society that pronounced Newton the sole inventor. What that investigation's readers were not told, at the time, was that Newton — as President of the Society — had written the report's conclusions himself, and later anonymously authored a favorable review of the report's own findings for publication.

This chapter presents Newton's own words from the dispute's later, most hardened years, without softening them, alongside the context needed to understand what he was actually claiming.

> *"Second inventors have no right. Whether Mr Leibniz found the Method by himself or not is not the Question. We take the proper question to be, who was the first inventor of the method."*
>
> — Newton's own account of the Commercium Epistolicum, published in the Philosophical Transactions, 1715

Writing anonymously in the third person about the very report he had authored, Newton lays out the governing principle of his entire case: independent discovery, however genuine, counted for nothing against the claim of the first inventor.

> *"To take away the Right of the first inventor, and divide it between him and that other, would be an Act of Injustice."*
>
> — Newton's own account of the Commercium Epistolicum, 1715

Newton frames the dispute in the language of property and theft rather than the more forgiving language of simultaneous discovery that he himself, earlier in the correspondence of the 1670s, had seemed willing to entertain.

It is worth setting this hardened, later Newton beside the younger man visible in the 1676 letters to Leibniz that opened their correspondence — cordial, even generous, sharing mathematical results across the English Channel with every sign of collegial respect. Something changed in the intervening decades, and historians still argue over how much of it was genuine grievance and how much was wounded pride, national rivalry between English and Continental mathematicians, or Newton's own well-documented difficulty ever letting a dispute over credit go unresolved in his own favor.

It is also worth being honest about what the anonymous authorship of the Royal Society's report actually means. Newton did not merely hold a strong private opinion about the dispute; he used his own institutional authority, as President of the very body conducting the investigation, to shape its conclusions while presenting those conclusions to the public as the impartial judgment of his peers. Measured against the standard this entire book has tried to hold him to — public claims resting only on what could actually be shown, kept honestly separate from private conviction — this episode is the clearest failure of that standard anywhere in his life. He did not feign a hypothesis here. He feigned impartiality.

Leibniz died in 1716, the dispute still unresolved between them, his own final years consumed by an argument that had, by most contemporary accounts, isolated him from former allies and darkened his reputation across Europe. Newton outlived him by more than a decade and continued revising the historical record of the dispute for years afterward. Modern scholarship's judgment — that both men independently and legitimately arrived at the same mathematics — was not available to either of them while they were alive to hear it, and it is worth remembering, closing this chapter, that the vindication history eventually offered Leibniz came far too late to do him any good.

This book has tried throughout to show a man whose discipline in the pursuit of evidence rarely wavered. This chapter is the clearest place that discipline gave way — not to bad science, but to something more ordinary and more human: a refusal, once wounded, to ever quite let go.

CHAPTER 24
Friendship

"Philosophy is such an impertinently litigious Lady that a man had as good be engaged in Law suits as have to do with her."

— Letter to Edmond Halley, 20 June 1686

Every rivalry examined so far in this book needs a counterweight, and this chapter is it. Edmond Halley — astronomer, mathematician, and later Astronomer Royal — was arguably more responsible than anyone besides Newton himself for the Principia's existence. It was Halley who, in August 1684, made the trip to Cambridge that started everything: he asked Newton, almost as an aside, what curve a planet would trace if pulled by a force weakening with the square of its distance from the sun. Newton answered immediately that it would be an ellipse, and told Halley he had already calculated it years before but could not, on the spot, find the paper. Halley pressed him to redo the work and write it up properly. The Principia, arguably the single most important scientific book ever published, grew directly out of that one conversation.

Halley then spent the better part of three years shepherding the book into existence: pushing Newton to expand a short paper into a full three-book treatise, managing his volatile moods through the Hooke dispute examined earlier in this book, correcting the proofs himself, designing the woodcut illustrations, and — when the Royal Society's own funds ran short, having just spent its publication budget on an expensive, poorly-selling book about fish — paying for the Principia's printing out of his own pocket. He also wrote the anonymous prefatory ode in

Newton's honor that opens the book, and drafted its first published review.

This chapter draws on the correspondence between the two men during the Principia's most difficult stretch, when Newton's dispute with Hooke nearly cost the book its most important section.

> *"The proof you sent me I like very well... The third I now design to suppress. Philosophy is such an impertinently litigious Lady that a man had as good be engaged in Law suits as have to do with her. I found it so formerly & now I no sooner come near her again but she gives me warning."*
>
> — Letter to Edmond Halley, 20 June 1686

At the height of the dispute with Hooke, Newton threatened to withhold Book III of the Principia entirely — the section containing the System of the World, the part of the book most readers would eventually find comprehensible and compelling. Halley's tact, patience, and refusal to simply accept the threat is what saved it.

> *"I wish I had spared ye Postscript in my last."*
>
> — Letter to Edmond Halley, 1686

Having lashed out at Hooke in an earlier letter's postscript, Newton — evidently talked down by Halley's careful, diplomatic reply — offers something rare in his correspondence: an unprompted admission that he had gone too far.

It would be easy to read the 'litigious Lady' line as simply more evidence of Newton's thin skin, and in one sense it is. But read beside everything else in this book, it also captures something true and worth taking seriously: Newton experienced the pursuit of natural philosophy as something closer to a demanding, occasionally punishing relationship than a tidy accumulation of facts. He had been burned by public dispute before — the 1672 reception of his color theory, examined early in this book — and the wariness in this letter is the wariness of a man who has learned, through direct experience, exactly how costly that kind of engagement could be.

What makes Halley's role remarkable is that he absorbed all of this — the threats, the reversals, the demands — without ever losing either his own temper or his commitment to getting the book published. He was not merely an editor or a patron; he functioned, across these critical months, as something closer to a translator between Newton's private volatility and the public, finished work the world eventually received. Few scientific collaborations in history have depended so completely on one man's patience absorbing another man's temperament.

It is worth noting, finally, that Newton never forgot the debt. Decades later, as President of the Royal Society, he supported Halley's own scientific ambitions and, unlike his treatment of Hooke, the relationship between the two men remained cordial for the rest of their lives. Whatever else can be said about Newton's difficulty sustaining friendships — and this book has not shied away from that difficulty elsewhere — his bond with Halley is the clearest surviving counter-example: a rivalry that never happened, precisely because one man was patient enough to prevent it.

CHAPTER 25

Money

"I caused a great many of them to be assayed in the Mint & found by the assays that fine Gold was to fine Silver in Spain, Portugal, France, Holland, Italy, Germany & and the Northern Kingdoms in the proportions above mentioned."

— Report to the Lords Commissioners of the Treasury, 21 September 1717

An earlier chapter in this book followed Newton the detective, tracking counterfeiters through the taverns and prisons of London. This chapter follows a different side of the same job: Newton the monetary theorist, using the same tools of careful measurement and cross-checked evidence to solve a problem that had quietly troubled England's currency for decades.

By the 1710s, English gold guinea coins were officially valued at a fixed rate against silver shillings — but the official rate did not match the rate silver and gold actually traded at across the rest of Europe. The mismatch created a simple, ruinous incentive: anyone could buy gold cheaply abroad, bring it to England, exchange it for an overvalued amount of silver coin, then melt that silver down and sell it abroad at its true value, pocketing the difference. Silver was steadily draining out of the country as a direct result, and by 1717 the Treasury asked its Master of the Mint to find out why, and what to do about it.

Newton's answer was not guesswork. He personally assayed samples of foreign coin passing through the Mint, and gathered exchange data from merchants trading with Spain, Portugal, France, Holland, Italy, Germany, and

Sweden, building an actual cross-national table of gold-to-silver ratios before drawing any conclusion at all.

> *"Coyns abounded in England, I caused a great many of them to be assayed in the Mint & found by the assays that fine Gold was to fine Silver in Spain, Portugal, France, Holland, Italy, Germany & and the Northern Kingdoms in the proportions above mentioned."*
>
> — Report to the Lords Commissioners of the Treasury, 21 September 1717

Newton's report opens not with a recommendation but with a description of his method — the same instinct traced across this entire book, applied here to foreign exchange rates instead of planetary orbits.

> *"It is the demand for Exportation which hath raised the price of Exportable Silver about 2d. or 3d. in the ounce above that of Silver in Coyn, & hath thereby created a Temptation to export or melt down the Silver Coyn rather than give 2d. or 3d. more for Forreign Silver."*
>
> — Report to the Lords Commissioners of the Treasury, 21 September 1717

Newton diagnoses the precise mechanism driving silver out of the country — a small, specific price gap, stated in pence per ounce, that made melting down English coin more profitable than buying silver honestly.

> *"And therefore may be diminished by lowering the value of Gold in proportion to Silver."*
>
> — Report to the Lords Commissioners of the Treasury, 21 September 1717

Having diagnosed the cause, Newton states his remedy in the same flat, unadorned prose: adjust the official ratio, and the incentive to melt down silver coin should shrink along with it.

Parliament acted on Newton's report within months, fixing the guinea's value at twenty-one shillings — a small, technical adjustment that historians now regard as the moment England was effectively, if not deliberately, set on course toward the gold standard: a monetary system that would eventually spread across much of the world and last, in various forms, for the better part of two centuries. Newton did not set out to reshape international finance. He set out to answer a specific, bounded question about why silver kept leaving the country, using exactly the same evidence-gathering discipline this book has traced through every other chapter, and the consequences simply outran the scope of the original problem.

It is worth noting how little theoretical apparatus this report actually required, compared to the mathematics of the Principia. There is no calculus here, no elegant proof — only patient measurement, cross-checked against multiple independent sources, and a plain statement of cause and remedy. It is, in its way, the most accessible piece of Newton's own writing collected anywhere in this book: recognizably the same mind, working on a problem any literate merchant of his day could, in principle, have followed line by line.

CHAPTER 26

Prophecy

"The folly of Interpreters has been to foretell times and things by this Prophecy, as if God designed to make them Prophets."

— Observations upon the Prophecies of Daniel, and the Apocalypse of St. John (1733)

This book has already drawn, carefully and in limited scope, on Newton's private theological manuscripts. This chapter turns to a different body of his religious writing: Observations upon the Prophecies of Daniel, and the Apocalypse of St. John, a full-length work Newton spent decades preparing and revising, published posthumously in 1733. Unlike the Trinity material examined earlier, this was writing Newton fully intended to eventually publish — he simply never finished refining it to his own satisfaction during his lifetime.

The book's governing idea is distinctive and worth stating plainly: Newton did not believe biblical prophecy was meant to let readers predict the future. He believed it followed a consistent internal symbolic language — that a beast, a horn, a set number of days meant roughly the same thing wherever they appeared across different prophetic books — and that this language could be studied and decoded systematically, the same way a natural philosopher might study the grammar of an unfamiliar phenomenon. Prediction, in his view, was not the point. Confirmation, after the fact, was.

> *"The folly of Interpreters has been to foretell times and things by this Prophecy, as if God designed to make them Prophets. By this*

> *rashness they have not only exposed themselves, but brought the Prophecy also into contempt."*
>
> — Observations upon the Prophecies of Daniel, and the Apocalypse of St. John (1733)

Newton opens with a direct rebuke of the very practice most people associate with biblical prophecy — using it to forecast future events — and states plainly that he considers this approach both presumptuous and self-defeating.

> *"For as the few and obscure Prophecies concerning Christ's first coming were for setting up the Christian religion, which all nations have since corrupted, so the many and clear Prophecies concerning the things to be done at Christ's second coming are not only for predicting but also for effecting a recovery and re-establishment of the long-lost truth."*
>
> — Observations upon the Prophecies of Daniel, and the Apocalypse of St. John (1733)

Newton connects his prophetic scholarship to the same conviction visible in the chapters on his private theology: that Christianity, as commonly practiced in his own time, had drifted from an original, purer form he believed careful study could help restore.

It would be a mistake to treat this project as entirely separate from the method examined in this book's opening chapters. Newton approaches Daniel and Revelation the way he approached a prism or a planetary orbit: gathering the available evidence — in this case, the text's own recurring imagery — before allowing himself to draw a general conclusion, and explicitly warning his readers against the kind of premature, unearned certainty he associated with failed prophetic predictions throughout history. The domain is entirely different. The discipline is recognizably the same.

This does not mean the project meets the same evidentiary standard as his physics, and this book will not claim that it does. A symbolic reading of ancient prophetic literature cannot be tested by experiment the way a claim about refracted light can. What can be said honestly is that Newton approached the text believing it followed real, discoverable rules rather than admitting of any interpretation a reader wished to impose — and that conviction, however contestable its foundation, is recognizably an extension of the same mind found everywhere else in this book, refusing to accept an easy answer where a harder, more careful one might be found instead.

CHAPTER 27

A Difficult Year

"I am extremely troubled by the embroilment I am in, have neither ate nor slept well in the last twelve months, nor have my former consistency of mind."

— Letter to Samuel Pepys, 13 September 1693

This book has, since its opening pages, drawn a firm line around the material it will use: nothing offered as Newton's own words without a document to show for it, and particular caution around anything touching his final years or death, where apocryphal material clusters most thickly. This chapter observes that same caution, applied now to a different period — the late summer and early autumn of 1693, when Newton, then fifty years old, wrote two letters that survive in his own hand describing a period of real personal distress.

This book will not attempt to diagnose what Newton experienced, then or now. What follows is confined to what he himself wrote, to two named correspondents, on dated letters that have never been seriously disputed by historians.

> *"I am extremely troubled by the embroilment I am in, have neither ate nor slept well in the last twelve months, nor have my former consistency of mind."*
>
> — Letter to Samuel Pepys, 13 September 1693

A plain, restrained description of sustained difficulty with sleep, appetite, and concentration, addressed to a friend rather than offered for any public record.

> *"Being of opinion that you endeavoured to embroil me with woemen & by other means I was so much affected with it as that when one told me you were sickly and would not live I answered twere better you were dead. I desire you to forgive me this uncharitableness."*
>
> — Letter to John Locke, September 1693

Newton apologizes, unprompted, for an accusation he had evidently made against Locke some time before — and, in the same breath, discloses the alarming thought that had accompanied it, without excusing it.

> *"I am your most humble and unfortunate servant, Is. Newton."*
>
> — Letter to John Locke, September 1693

The closing of the same letter — a sign-off strikingly different from Newton's usual correspondence, where he typically signed himself simply as a servant or friend, without the added word 'unfortunate.'

Locke's reply, which survives, is notably generous: he accepts the apology without apparent rancor and asks only that Newton point out specifically what in Locke's own writing had given the earlier offense, so that he might address it directly. Whatever passed between the two men, their correspondence continued afterward, and no lasting rupture in the friendship is recorded.

Historians have speculated for a century about what lay behind this period — exhaustion from the years spent finishing the Principia, a disrupted friendship with the Swiss mathematician Nicolas Fatio de Duillier that ended around the same time, even, more speculatively, the cumulative effect of decades spent handling mercury and other substances in his alchemical experiments. This book has deliberately not adjudicated between these theories, for the same reason it has declined, throughout, to fill any other gap in the record with a confident guess dressed up as fact. What can be said is only what Newton himself set down: a difficult year, honestly described, to two men he trusted enough to tell.

It is worth closing this chapter by noting what came after. Newton did not vanish from view, and no lasting professional consequence followed. Within three years he had accepted the Wardenship of the Mint and begun the second half of his working life described elsewhere in this book — the Great Recoinage, the pursuit of counterfeiters, three decades as President of the Royal Society. Whatever the autumn of 1693 was, it was not an ending. The same discipline that examined a prism, a planet, and a coiner's confession also, on the strength of this brief and honest correspondence, seems to have eventually turned back toward steadier ground.

Notes and Sources

Every passage in this book attributed to Isaac Newton is drawn from a document that survives in his own hand or in an authoritative early printed edition: the Opticks (1704, revised through the fourth edition of 1730), the Principia Mathematica (1687, with later revisions), his correspondence — much of it now held and transcribed by the Newton Project at the University of Oxford — the Mint Papers from his three decades at the Royal Mint, and his theological and chronological writings, including The Chronology of Ancient Kingdoms Amended and Observations upon the Prophecies of Daniel and the Apocalypse of St. John.

Three widely repeated Newton quotations were deliberately excluded from this book because they cannot be reliably traced to anything Newton himself wrote or said:

— The image of a boy playing on a seashore, distracted by a smooth pebble or a prettier shell while the great ocean of truth lay undiscovered before him. First published more than a century after Newton's death, and attributed secondhand to a source who was not in England at the time the remark was supposedly made.

— The claim that Newton "thinks half-heartedly" about ordinary matters. No traceable original exists for this paraphrase.

— A remark comparing obstacles to walls or bridges, meant merely to be climbed, gone through, or gone around. This, too, cannot be traced to any surviving Newton document.

Where a chapter draws on a quotation whose precise dating or attribution is disputed among historians, that uncertainty is noted directly in the text rather than smoothed over.

This is the first of what is intended as a broader look at Newton's life in his own words. Readers interested in the primary sources behind this book are encouraged to consult the Newton Project (newtonproject.ox.ac.uk), which continues to publish newly transcribed manuscripts from Newton's surviving papers.

Shumba

Shumba is the Shona word for lion. This name was given to Ian Christopher Thorpe by Paramount Chief Abisha Mutasa when Ian was 18 years old. Abisha's kingdom included most of Zimbabwe and part of Mozambique. At the time, they were standing on a bridge over Odzi River that Ian had built together with the children from St Matthias High School where he was a teacher.

Ian went on to found and lead The Africa Trust. He invented the award-winning Elephant Pump and improved approaches for sanitation. He leads the team which has so far enabled over five million people to gain access to sustainable supplies of clean water and decent sanitation. A hospital and pharmacy have been built along with bridges. Thousands of village banks have been established, allowing people to save, borrow and lend among themselves.

Entrepreneurial training with loans has helped over 50,000 people start their own businesses in Malawi, Zimbabwe, Tanzania, Kenya and Uganda. In Buduburam Refugee Camp in Ghana, Ian heard heart-wrenching testimonies from 90 Liberian former child soldiers, then in their twenties. He helped to reintegrate them back into their home villages in Liberia by teaching them how to build pumps and toilets after they had been expelled from the camp and forced to return home.

Part of the profit from this series of books will go towards this philanthropic work that is so close to Ian's heart.